Fundamentals of
ELECTROWEAK THEORY

Fundamentals of ELECTROWEAK THEORY

by

Tarun Chada | Dhruv Aggarwal

Wordpen Academics
Publishing Science That Matter

ISBN: 978-93-6084-662-6

Published by : **WORDPEN ACADEMICS**
4736/23, Ansari Road, Darya Ganj,
New Delhi -110 002
Phone: +91-011-35004336
E-mail: info@wordpenacademics.com

Printed at **:** **Replika Press Pvt. Ltd.**

Exclusive Distributor: ***Bio-Green Books, New Delhi***

Preface

"Fundamentals of Electroweak Theory" delves into the profound and elegant framework that underpins our understanding of the fundamental forces governing the subatomic realm. This book is a testament to the ingenuity and collaborative efforts of generations of physicists who have unraveled the mysteries of the electroweak interaction, laying the groundwork for our modern understanding of particle physics.

The electroweak theory represents one of the crowning achievements of theoretical physics, unifying the electromagnetic and weak nuclear forces within a single, coherent framework. At its heart lies the Higgs mechanism, which endows particles with mass and provides crucial insights into the nature of symmetry breaking and the origin of mass in the universe.

As editors of this volume, we have endeavored to compile a comprehensive resource that introduces readers to the foundational principles, mathematical formalism, and experimental evidence underlying the electroweak theory. Through a blend of theoretical concepts, experimental results, and computational techniques, this book provides a thorough understanding of the structure and predictions of the electroweak theory, as well as its implications for particle physics and cosmology.

Our overarching goal in presenting this book is to inspire curiosity, foster deeper insights, and stimulate further inquiry into the fascinating world of particle physics. By demystifying complex theoretical concepts and showcasing the experimental evidence that supports them, we aim to empower students, researchers, and enthusiasts to engage with the frontiers of knowledge and contribute to the ongoing quest for a deeper understanding of the fundamental laws of nature.

We extend our sincere gratitude to all the contributors who have generously shared their expertise, insights, and enthusiasm for the electroweak theory, as well as to the readers who embark on this intellectual journey with us. May this volume serve as a valuable resource and inspiration for all those who seek to unravel the mysteries of the universe through the lens of electroweak theory.

Tarun Chada | Dhruv Aggarwal

Contents

Part I

The road to unification

1

The electromagnetic current and its properties

1.1 Introduction

The theory of the weak interactions, better known as the electroweak theory, was developed in two stages. In the first stage, a phenomenological interaction was introduced and was extended when additional experimental results became available. At that stage a large number of observations could be accounted for by empirical rules. There still remained the desire to develop a basic theory that was finite and renormalizable. This was achieved in the second stage by combining the electromagnetic and weak interactions into a gauge theory – the electroweak theory.

The effective current–current interaction was introduced by Fermi in 1934,

$$\mathcal{H}_{\text{eff}} = -\frac{G_{\text{F}}}{\sqrt{2}} J_\mu(x) J^{\mu\dagger}(x), \tag{1.1}$$

and was responsible for charged-current weak interactions of leptons and hadrons. The current was originally introduced, in analogy to electrodynamics, for the interaction of the electron with its neutrino and also for the neutron–proton transition

$$J_\mu(x) = \overline{\Psi}_{\nu_e}\gamma_\mu(1-\gamma_5)\Psi_{\text{e}} + \overline{\Psi}_{\text{p}}\gamma_\mu(1-\gamma_5)\Psi_{\text{n}} + \overline{\Psi}_{\nu_\mu}\gamma_\mu(1-\gamma_5)\Psi_\mu + \cdots. \tag{1.2}$$

Here the Ψs are the fields of the fermions and the γs are the Dirac γ-matrices in the notation of Bjorken and Drell (1965). The shortcoming of this theory is known as the unitarity problem and shows up in many reactions. For example, for the reaction

$$\nu_\mu + \text{e}^- \longrightarrow \nu_{\text{e}} + \mu^-$$

we can calculate the cross section, which to lowest order is

$$\sigma_{\text{tot}}(\nu_\mu \text{e}^- \to \nu_{\text{e}}\mu^-) = \frac{G_{\text{F}}^2 s}{\pi} \tag{1.3}$$

with $s = 4E_{\rm cm}^2$, where terms proportional to the masses of the leptons have been omitted at high energies. Because of the point coupling in (1.1) only the lowest partial wave (angular momentum zero) can contribute to the scattering amplitude. Then conservation of probability (unitarity) in quantum mechanics requires (see Problems 1 and 2 at the end of Chapter 2)

$$\sigma_{\rm inelastic}^{l=0} \leq \frac{\pi}{2E_{\rm cm}^2} \tag{1.4}$$

for any scattering process. From (1.3) and (1.4) we find that the theory is consistent with unitarity only for

$$E_{\rm cm} \leq \left(\frac{\pi\sqrt{2}}{4G_{\rm F}}\right)^{\frac{1}{2}} = 309\,{\rm GeV}. \tag{1.5}$$

Thus the theory is incomplete.

On the other hand, why should we believe the first-order-term result for such high energies? It is not a matter of belief but an unfortunate fact of life that we cannot calculate higher-order contributions. The theory, which is based on the Hamiltonian (1.1), is non-renormalizable and does not allow a well-defined perturbation expansion.

At this point we fall back upon the most successful field theory at our disposal: quantum electrodynamics (QED). We describe in this chapter its salient features and we try to develop in Part II of this book, in analogy to QED, a gauge theory of weak and electromagnetic interactions. In fact the second stage in the development of the weak interactions is to construct a well-defined and renormalizable theory.

We start with the Dirac Lagrangian for an electron interacting with the electromagnetic field,

$$\mathcal{L} = \overline{\Psi}\left(\mathrm{i}\gamma^\mu \frac{\partial}{\partial x^\mu} + e\gamma^\mu A_\mu - m\right)\Psi - \frac{1}{4}F_{\mu\nu}F^{\mu\nu}. \tag{1.6}$$

We think of Ψ as the electron field whose current

$$j_\mu = \overline{\Psi}(x)\gamma_\mu\Psi(x) \tag{1.7}$$

interacts with the electromagnetic field

$$\mathcal{L}_{\rm F} = \overline{\Psi}(\mathrm{i}\gamma^\mu\partial_\mu - m)\Psi + ej^\mu A_\mu. \tag{1.8}$$

The interaction term $e\overline{\Psi}\gamma_\mu\Psi A^\mu$ fixes the vertex and the electron propagator is the inverse of the kinetic term.

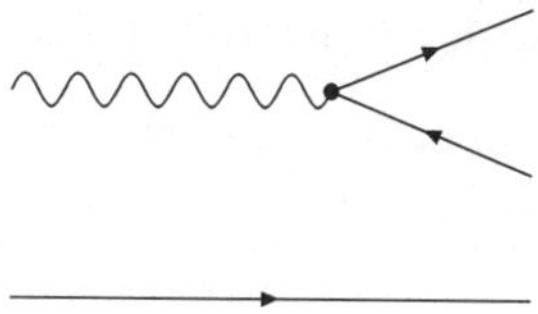

$ie\gamma_\mu$

$\frac{i}{\not{p} - m + i\varepsilon}$

Figure 1.1. The photon–fermion vertex and the propagator.

Finally, the last term in (1.6) gives the interaction between photons and involves the electromagnetic field tensor

$$F_{\mu\nu} = \frac{\partial A_\nu}{\partial x^\mu} - \frac{\partial A_\mu}{\partial x^\nu}. \tag{1.9}$$

Gauge invariance forbids a term $m_\gamma^2 A_\mu A^\mu$ that would give a mass to the photon. QED has been one of the most precise and successful theories in all of physics and has been tested to a few parts per million.

As mentioned above, the electromagnetic current describes the interaction of the photon with a charged fermion. The current is a local operator

$$j_\mu(x) = \overline{\Psi}_l(x)\gamma_\mu\Psi_l(x), \tag{1.10}$$

where $\Psi_l(x)$ is the field for the lepton l and γ_μ is a Dirac matrix. The current $j_\mu(x)$ is a generalization of the classical concept of a current as it appears in Maxwell's theory. In classical electrodynamics $j_\mu(x)$ is a four-vector with components

$$\begin{aligned} j^\mu(x) &= \left[c\rho(x), \vec{j}(x) = \rho(x)\vec{v}\right] \\ &= \rho(x)\left[c, \vec{v}\,\right], \end{aligned} \tag{1.11}$$

with $\rho(x)$ denoting the charge density, the vector $\vec{j}(x)$ the charge flow, c the speed of light, and $\vec{v}$ the velocity of the charge density. The total charge of a particle is given by the integral

$$c\,Q = \int \mathrm{d}^3x\, j_0(x). \tag{1.12}$$

The current in (1.11) is an operator that transforms like a four-vector. The fields occurring above are also operators that create and destroy localized particle states. They satisfy canonical commutation relations, which quantize the theory. The computational methods of QED can be found in many books given in the references. We shall assume that the reader is familiar with the methods of quantum electrodynamics.

1.2 The current for hadronic states

The electromagnetic current for a proton is more complicated since protons are not point-like particles, but have a measurable physical size formed by the cloud of pions and other hadrons which surrounds them. As a first attempt one would write the electromagnetic current for a proton in terms of free fields,

$$J_\mu = \overline{\Psi}_{\mathrm{p}'}(x)\gamma_\mu \Psi_{\mathrm{p}}(x) = \overline{u(p')}\gamma_\mu u(p)\mathrm{e}^{\mathrm{i}(p'-p)x}. \tag{1.13}$$

This form is ruled out immediately because it describes a point particle with unit charge and a Dirac magnetic moment. It obviously fails for the case of a proton, which has size and an anomalous magnetic moment. This implies a charge distribution and requires additional terms on the right-hand side.

One therefore expects a more general structure, which is introduced by considering the hadronic current as a vector operator that satisfies general symmetry principles. We begin by considering the interaction of the electromagnetic field $A_\mu(x)$ with protons. The matrix element contains the term

$$\langle p'|J_\mu(x)\mathrm{e}^{-\mathrm{i}qx}|p\rangle. \tag{1.14}$$

Under translations in space and time $J_\mu(x)$ transforms as

$$J_\mu(x) = \mathrm{e}^{\mathrm{i}\hat{p}x} J_\mu(0)\mathrm{e}^{-\mathrm{i}\hat{p}x}, \tag{1.15}$$

where $\hat{p}$ is the operator of the four-momentum; thus (1.14) reduces to

$$\int \mathrm{d}^4x \langle p'|J_\mu(0)|p\rangle \mathrm{e}^{-\mathrm{i}(q+p-p')x} = \bar{u}(p')O_\mu(p',p)u(p) \int \mathrm{d}^4x\, \mathrm{e}^{-\mathrm{i}(q+p-p')x}, \tag{1.16}$$

with O_μ containing terms with γ-matrices, the antisymmetric tensor $\varepsilon_{\mu\nu\alpha\beta}$, and momenta. The spinors $u(p)$ and $u(p')$ are solutions of the free Dirac equation. These are the requirements of Lorentz invariance.

Two other properties are

(i) gauge invariance, which translates into

$$q^\mu \langle p'|J_\mu(0)|p\rangle = q^\mu \bar{u}(p')0_\mu u(p) = 0; \tag{1.17}$$

(ii) Hermiticity of the current

$$\begin{aligned} \langle p'|J_\mu(0)|p\rangle^* &= \langle p|J_\mu(0)|p'\rangle, \\ \left[\bar{u}(p')0_\mu u(p)\right]^+ &= \bar{u}(p)0_\mu u(p'), \end{aligned} \tag{1.18}$$

from which it follows that

$$O_\mu^+ = \gamma_0 O_\mu \gamma_0. \tag{1.19}$$

The above requirements limit the types of Dirac matrices and momenta which are included in the operator O_μ. The first subset of operators is

$$\{\ell_\mu = p_\mu + p'_\mu,\ q_\mu = p'_\mu - p_\mu,\ \gamma_\mu,\ \mathrm{i}\sigma_{\mu\nu}q^\nu,\ \sigma_{\mu\nu}\ell^\nu\}, \tag{1.20}$$

which appear in matrix elements of the vector current. In addition to the above operators, there are also others that contain γ_5 or the antisymmetric tensor. They are produced by higher-order weak or new interactions and their contributions to electromagnetic matrix elements are small. For completeness we include them here and discuss some properties in the next section. The second subset of matrices contains γ_5,

$$\{\ell_\mu\gamma_5,\ q_\mu\gamma_5,\ \gamma_\mu\gamma_5,\ \sigma_{\mu\nu}q^\nu\gamma_5\}, \tag{1.21}$$

and the third the antisymmetric tensor,

$$\{\varepsilon^{\mu\nu\alpha\beta}\sigma_{\alpha\beta}\ell_\nu,\ \varepsilon^{\mu\nu\alpha\beta}\sigma_{\alpha\beta}q_\nu,\ \varepsilon^{\mu\nu\alpha\beta}\gamma_\mu\gamma_5 q_\alpha\ell_\beta,\ \varepsilon^{\mu\nu\alpha\beta}\gamma_\nu q_\alpha\ell_\beta\}. \tag{1.22}$$

These terms are odd under parity transformations. Matrix elements of these operators are not all linearly independent. For instance, matrix elements of the last three terms in (1.22) are reduced, by judicious use of γ-matrix identities and the Dirac equation, to matrix elements of the second set (Nowakowski *et al.*, 2005).

The Gordon decomposition formula

$$\bar{u}(p')\gamma^\mu u(p) = \bar{u}(p')\left(\frac{p'^\mu + p^\mu}{2m} + \frac{\mathrm{i}\sigma^{\mu\nu}q^\nu}{2m}\right)u(p) \tag{1.23}$$

eliminates one term in the first subset. Similarly, the term $\sigma_{\mu\nu}\ell^\nu$ can be replaced by $\bar{u}(p')q_\mu u(p)$. Thus the matrix element of the vector current has the general form

$$\bar{u}(p')\left(\gamma_\mu F_1(q^2) + \frac{\mathrm{i}\sigma_{\mu\nu}q^\nu}{2m}F_2(q^2) + q_\mu F_3(q^2)\right)u(p). \tag{1.24}$$

Gauge invariance gives an additional condition,

$$F_3(q^2) = 0. \tag{1.25}$$

The functions F_i with $i = 1, 2, 3$ are Lorentz scalars and their argument must remain unchanged under the replacement $p_\mu \to p_\mu + k_\mu$ and $p'_\mu \to p'_\mu + k_\mu$ with k_μ an arbitrary four-vector; consequently they are functions of $q^2 = (p' - p)^2$, which justifies the argument introduced in Eq. (1.24). We can use the Hermiticity condition as written in (1.19) to assure that the form factors are real functions. In summary, symmetry principles restrict the number and properties of the form factors. Some other consequences of symmetries are discussed in Chapter 2 and the problems given there.

What is the physical meaning of form factors? As the name indicates, they describe the structure or configuration of particles. Let us begin with an electron in the Dirac theory. To lowest order of electrodynamics $F_1(0) = 1$ and $F_2(0) = 0$. On replacing next the γ_μ term with the help of the Gordon decomposition, the coupling of the electron to the electromagnetic field $A_\mu(x)$ is written as

$$e\bar{\Psi}_{\mathrm{f}}(x)\gamma_\mu A^\mu(x)\Psi_i(x) = e\bar{\Psi}_{\mathrm{f}}(x)\left(\frac{p_\mu + p'_\mu}{2m} + \frac{i\sigma_{\mu\nu}q^\nu}{2m}\right)\Psi_i(x)A^\mu(x). \quad (1.26)$$

The non-relativistic limit produces two terms. The first term, from the sum of momenta,

$$e\bar{u}(p')u(p)A^0(x) \quad (1.27)$$

couples the charge density to the scalar potential because the ratio of the three-momentum to the mass becomes very small. The second term couples the magnetic moment to an external magnetic field. Considering a constant magnetic field $\vec{B}$ and its potential $A_\mu(x)$, the interaction in configuration space is

$$\frac{e}{2m}\bar{\Psi}_{\mathrm{f}}(x)\sigma_{\mu\nu}\Psi_i(x)\,\frac{\partial A_\mu(x)}{\partial x^\nu} = \bar{\Psi}_{A,\mathrm{f}}(x)\frac{e}{2m}\vec{\sigma}\cdot\vec{B}\,\Psi_{A,i}, \quad (1.28)$$

where Ψ_A are the upper components of the spinors (see Problem 2.5). The magnetic field is introduced as the rotation of the vector potential. Defining the magnetic moment as

$$\vec{\mu} = -g\frac{e}{2m}\vec{S} \quad \text{with} \quad \vec{S} = \frac{\vec{\sigma}}{2}, \quad (1.29)$$

we obtain for the electron the gyromagnetic ratio $g = 2$. Thus a Dirac electron has an intrinsic magnetic moment with the natural value of 2, which can be modified by radiative corrections.

Although we have started to derive a current for extended fermions, the results of this derivation in the form of Eqs. (1.27) and (1.29) are also valid for "point-like" particles, when higher-order electromagnetic corrections are taken into account. Indeed, the Lagrangian given in (1.6) will induce correction terms compatible with the symmetries of the Lagrangian. We see from (1.23) and (1.24) that F_2 will also contribute to the magnetic moment via $\mu = \frac{1}{2}[F_1(0) + F_2(0)]$. Both for the electron and for the muon, the magnetic moments have been measured very accurately. They have also been calculated theoretically and the agreement is very good. For the electron

$$\frac{1}{2}(g-2)_{\mathrm{e}} = 0.001\,159\,652\,209\,(31), \quad (1.30)$$

with the number in parentheses denoting the experimental accuracy. Very accurate results exist also for the muons. The deviation from the value of 2 comes from radiative corrections, which in quantum electrodynamics have been calculated precisely (Kinoshita, 1990).

The situation is very different for protons and neutrons. The experimental values are

$$F_1(0) = 1 \quad \text{and} \quad F_2(0) = 1.79 \quad \text{for the proton,} \tag{1.31}$$

$$F_1(0) = 0 \quad \text{and} \quad F_2(0) = -1.91 \quad \text{for the neutron.} \tag{1.32}$$

The changes come from the strong interactions and cannot yet be calculated. They are called the anomalous magnetic moments and have been measured in electron–hadron-scattering experiments. In addition to their values at $q^2 = 0$, the form factors have been measured over extended regions of the momentum-transfer squared and were found to decrease rapidly with q^2. This behavior indicates the existence of a charge distribution of virtual particles around the proton and the neutron, with the charge density decreasing rapidly with increasing radius. The motion of the particles creates magnetic fields, which are manifested in the values of the magnetic moments.

1.3 Parity-violating form factors

For completeness we include additional couplings of the photon induced by weak interactions inside the vertex. Omitting this section will not affect the study of the following chapters.

The electromagnetic force is not the only force between particles. For instance, the presence of weak terms changes the general structure of the electromagnetic matrix elements. The interaction of a photon with a particle does not mean that the whole process is electromagnetic, since higher-order corrections must also include the weak interactions. Conceptually it is easy to include these effects in the electromagnetic current, by dropping the restrictions that the current is invariant under the discrete symmetries charge conjugation C, parity P, and time-reversal T. Imposing Lorentz invariance, gauge invariance, and Hermiticity means that one must include two additional form factors (F_3 and F_4) and the electromagnetic current takes a more general form,

$$\begin{aligned}\bar{u}(p')0_\mu u(p) = \bar{u}(p')\Bigg[&\gamma_\mu F_1(q^2) + \mathrm{i}\frac{\sigma_{\mu\nu}q^\nu}{2m}F_2(q^2) + \mathrm{i}\frac{\epsilon_{\mu\nu\alpha\beta}\sigma^{\alpha\beta}q^\nu}{4m}F_3(q^2) \\ &+ \left(q_\mu - \frac{q^2}{2m}\gamma_\mu\right)\gamma_5 F_4(q^2)\Bigg]u(p).\end{aligned} \tag{1.33}$$

We know from classical electrodynamics and quantum mechanics that the fields transform under parity P and time-reversal T as shown in Table 1.1.

Table 1.1

$\vec{B}$	$\xrightarrow{\text{P}}$	$\vec{B}$
$\vec{B}$	$\xrightarrow{\text{T}}$	$-\vec{B}$
$\vec{E}$	$\xrightarrow{\text{P}}$	$-\vec{E}$
$\vec{E}$	$\xrightarrow{\text{T}}$	$\vec{E}$
$\vec{\sigma}$	$\xrightarrow{\text{P}}$	$\vec{\sigma}$
$\vec{\sigma}$	$\xrightarrow{\text{T}}$	$-\vec{\sigma}$

From Table 1.1 we can infer immediately that $\vec{\sigma} \cdot \vec{B}$, an interaction defining the second form factor $F_2(q^2)$, conserves parity and time-reversal. Similarly, the non-relativistic reduction of all form factors including $F_3(q^2)$ and $F_4(q^2)$ is given by

$$\mathcal{H}_{\text{int}} \propto eA_0 - \mu\vec{\sigma} \cdot \vec{B} - d\vec{\sigma} \cdot \vec{E} - a\left[\vec{\sigma} \cdot \left(\vec{\nabla} \times \vec{B} - \frac{\partial \vec{E}}{\partial t}\right)\right], \tag{1.34}$$

with $F_1(0) = e$ (charge), $[F_1(0) + F_2(0)]/(2m) = \mu$ (magnetic dipole moment), $F_3(0)/(2m) = d$ (electric dipole moment), and $F_4(0) \propto a$ is called the anapole moment (Zeldovich, 1958). It is evident that the presence of F_3 leads to a parity- and time-reversal-violating interaction. Physical phenomena that exhibit violation of time-reversal are very scarce. Therefore, the observation of $d \neq 0$ will be a physical breakthrough. Up to now only upper limits for d have been established for electrons and nucleons.

The fourth form factor $F_4(q^2)$ is even under time-reversal but violates parity. It is frequently omitted from discussions of the electromagnetic form factors, because it is an off-shell form factor, in the sense that its interaction with an on-shell photon vanishes. This is easily seen because $q^2 = 0$ and $\varepsilon_\mu q^\mu = 0$ for on-shell photons. In addition, this form factor can appear only in matter with currents producing the electromagnetic fields, because for classical fields the expression $\vec{\nabla} \times \vec{B} - \partial\vec{E}/\partial t$, which appears in the anapole interaction, vanishes (Maxwell equation in vacuum) in the absence of a current. Finally, for neutral fermions, which do not carry any global quantum numbers, like Majorana neutrinos, only the anapole form factor is possible. For a more detailed treatment of the form factors I recommend Nowakowski *et al.* (2005).

References

Bjorken, J. D., and Drell, S. D. (1965), *Relativistic Quantum Mechanics* (New York, McGraw-Hill)

Kinoshita, T. (1990), *Quantum Electrodynamics* (Singapore, World Scientific)

Nowakowski, M., Paschos, E. A., and Rodriguez, J. M. (2005), *Eur. J. Phys.* **26**, 545
Zel'dovich, Ya. B. (1958), *Sov. Phys. JETP* **6**, 1184

Select bibliography

Chapters 1–3 discuss various aspects in the early development of weak interactions. The following books define the notation (with small differences) that we shall use and include details on the early theory of weak interactions.

Bjorken, J. D., and Drell, S. D. (1965), *Relativistic Quantum Mechanics* (New York, McGraw Hill)
Gasiorowicz, S. (1966), *Elementary Particle Physics* (New York, John Wiley and Sons)
Lee, T. D. (1981), *Particle Physics and Introduction to Field Theory* (New York, Harwood Academic Publishers)
Marshak, R. E., Riazuddin, and Ryan, C. P. (1969), *Theory of Weak Interactions in Particle Physics* (New York, Interscience)
Nachtmann, O. (1990), *Elementary Particle Physics. Concepts and Phenomena* (Berlin, Springer-Verlag)
Peskin, M. E., and Schroeder, D. V. (1995), *Quantum Field Theory* (Reading, MA, Perseus Books)

An introductory presentation to various topics of this book appears in

Halzen, F., and Martin, A. D. (1984), *Quarks and Leptons* (New York, John Wiley and Sons)

2

The weak currents

2.1 The weak currents and some of their properties

The effective weak interaction in Eq. (1.1) was motivated by nuclear β-decays. For many years this was the main theoretical framework for analyzing experiments. As new experimental discoveries became available, the form of the interaction was maintained but the current $J_\mu(x)$ was enlarged to incorporate the new observations. At the end of the sixties the charged current $J_\mu^\dagger(x)$ included a leptonic and a hadronic term,

$$J_\mu^\dagger = l_\mu^\dagger(x) + h_\mu^\dagger(x). \tag{2.1}$$

The leptonic part of the current is

$$l_\mu^\dagger(x) = \overline{\Psi}_e(x)\gamma_\mu(1-\gamma_5)\Psi_{\nu_e}(x) + \overline{\Psi}_\mu(x)\gamma_\mu(1-\gamma_5)\Psi_{\nu_\mu}(x), \tag{2.2}$$

with the first term corresponding to the electron and its neutrino and the second term to the muon and its neutrino. Its space-time structure has a vector part analogous to the electromagnetic current and an axial part introduced after the discovery of parity violation. A direct calculation using the currents in (2.2) gives the μ-decay spectrum, which is in good agreement with experiment. It also gives the decay rate of the muon as

$$\Gamma(\mu \to e + \nu_e + \bar{\nu}_\mu) = \frac{G_\mu m_\mu^5}{192\pi^3}. \tag{2.3}$$

From the observed decay rate and the mass of the muon the constant G_μ is determined to be

$$G_\mu = (1.166\,32 \pm 0.000\,04) \times 10^{-5}\,\mathrm{GeV}^{-2}. \tag{2.4}$$

This determination includes the effects of radiative corrections, which in the electroweak theory are finite and can be calculated precisely.

The hadronic current consists of several parts determined by detailed analyses of hadron decays. For instance, the decay of a neutron, $n \to p + e^- + \nu_e$, is well described by the matrix element

$$\langle p|J_\mu^\dagger|n\rangle = \langle p|V_\mu^\dagger|n\rangle - \langle p|A_\mu^\dagger|n\rangle, \tag{2.5}$$

where we can decompose the matrix elements in terms of form factors. Lorentz invariance gives the general expressions

$$\langle p|V_\mu^\dagger|n\rangle = \bar{u}(p')\left(g_V\gamma_\mu + f_V\frac{(p+p')_\mu}{2M} + h_V\frac{q_\mu}{2M}\right)u(p) \tag{2.6}$$

and

$$\langle p|A_\mu^\dagger|n\rangle = \bar{u}(p')\left(g_A\gamma_\mu\gamma_5 + f_A\frac{\mathrm{i}\sigma_{\mu\nu}q^\nu}{2M}\gamma_5 + h_A\frac{q_\mu}{2M}\gamma_5\right)u(p), \tag{2.7}$$

where p_μ and p'_μ are the momenta of the neutron and proton, respectively, with $q_\mu = p'_\mu - p_\mu$. The functions g_V, f_V, and h_V are vector form factors describing the effects of strong interactions in the hadrons. Similarly g_A, f_A, and h_A are axial form factors. General symmetries, like charge symmetry and time-reversal, limit the form factors and demand that $h_V = f_A = 0$ (see Marshak *et al.*, 1969, p. 314). At zero momentum transfer, the vector form factor g_V was precisely determined and it is strikingly close to 1, while g_A is about -1.23. An explanation was proposed, namely that the strangeness-conserving part of $V_\mu^\dagger$ has the isospin content

$$V_\mu^\dagger = V_\mu^1 + \mathrm{i}V_\mu^2 =: V_\mu^+ \quad \text{and} \quad A_\mu^\dagger = A_\mu^1 + \mathrm{i}A_\mu^2 =: A_\mu^+, \tag{2.8}$$

where 1 and 2 denote the first and second components of isospin. This means that the charges

$$T^i = \int V_0^i(x)\mathrm{d}^3x \tag{2.9}$$

are the same isospin generators as those occurring in the strong interactions and are therefore conserved. This rule is called the conserved-vector-current (CVC) hypothesis. The T^i form an algebra that closes under commutation relations

$$\left[T^i, T^j\right] = \mathrm{i}\varepsilon^{ijk}T_k. \tag{2.10}$$

As a consequence, the commutator of T^+ with T^- produces the third component of isospin. In the late sixties T^3 had not yet been observed to mediate transitions with the strength G; there was no weak neutral current. But such an operator already existed in the electromagnetic current. The electromagnetic current consisted of two parts,

$$J_\mu^{\rm em}(x) = V_\mu^3(x) + \frac{1}{\sqrt{3}} V_\mu^8(x), \tag{2.11}$$

with $V_\mu^3(x)$ being the third component of isospin and V_μ^8 an iso-scalar current transforming as the eighth component of SU(3). It is evident that there is a relation between the weak and the electromagnetic currents, since the vector part of the weak current and the isovector part of the electromagnetic current form an isotriplet. The form of the interaction in (1.1) defines a universal coupling for leptonic, semi-leptonic, and non-leptonic decays. Once the coupling constant G has been determined, as in (2.4) from the muon decay, it can be used to translate the isotriplet hypothesis into relations between electromagnetic and weak matrix elements. In Section 11.3 we give a consequence of the isotriplet hypothesis and the cross section for neutrino–neutron quasi-elastic scattering. Expressions for the currents in terms of quark fields are given in Chapter 3.

Since V_μ^+ is an isospin current, its matrix elements at zero momentum transfer are simply given by Clebsch–Gordan coefficients. The strength $g_{\rm V}$ is determined in nuclear β-decay as well as in the elementary decays

$$\begin{aligned} \pi^+ &\longrightarrow \pi^0 + e^+ + \nu_e, \\ n &\longrightarrow p + e^- + \bar{\nu}_e. \end{aligned} \tag{2.12}$$

In all these cases the charge current connects states with the same isospin T, but different components T_3. At zero momentum transfer the relevant matrix element is

$$\langle I, I_3 + 1 | V_\mu^+(0) | I, I_3 \rangle = 1, \tag{2.13}$$

for $I = \frac{1}{2}$. The value of $g_{\rm V}$ is extracted from β-decay and its value is found (see Equation (9.28)) to be

$$g_{\rm V} = 0.9740 \pm 0.0003 \pm 0.0015. \tag{2.14}$$

This precise value includes radiative corrections, so its deviation from unity is significant.

Is the small difference of 2.6% a drawback of the theory or is there another component of the current? The discrepancy was explained by the observation that the hadronic current $V_\mu^\pm$ does not generate only the isospin group SU(2), but contains other pieces responsible for strangeness-changing decays, like

$$\begin{aligned} \Lambda^0 &\longrightarrow p + e^- + \bar{\nu}_e, \\ K^+ &\longrightarrow \pi^0 + e^+ + \nu_e. \end{aligned} \tag{2.15}$$

Thus the hadronic current is the sum of a $\Delta S = 0$ term and a $\Delta S = 1$ term,

$$V_\mu^+ = \cos\theta_c \, V_\mu^{\Delta S=0} + \sin\theta_c \, V_\mu^{\Delta S=1}. \tag{2.16}$$

The two terms are interpreted as two components of the current orthogonal to each other and connected through the mixing angle θ_c. The first term contains the isospin current that appears in (2.8),

$$V_\mu^{\Delta S=0} = V_\mu^1 + \mathrm{i}V_\mu^2. \tag{2.17}$$

The second component produces strangeness-changing transitions and it has, in SU(3), the form

$$V_\mu^{\Delta S=1} = V_\mu^4 + \mathrm{i}V_\mu^5. \tag{2.18}$$

The matrix elements of $V_\mu^{\Delta S=0}$ and $V_\mu^{\Delta S=1}$ can be estimated accurately. The conclusion from numerous experimental estimates gives the mixing angle

$$\sin\theta_c = 0.220 \pm 0.002. \tag{2.19}$$

In this way universality is restored, since the sum of the squares of the hadronic couplings reproduces the coupling observed in muon decay. In addition the discrepancy of g_V from 1 is understood. The angle θ_c is called the Cabibbo angle. The appearance of the Cabbibo angle will become more evident in the context of the Cabibbo–Kobayashi–Maskawa matrix, which enters the full Lagrangian of the weak interaction.

Finally, we mention one more difference between electromagnetic and weak interactions. The electromagnetic amplitude for the reaction $e^+e^- \to \mu^+\mu^-$ has the amplitude

$$\mathcal{M} = \mathrm{i}e^2 J^\mu \frac{g_{\mu\nu}}{q^2} J^{\nu\dagger}, \tag{2.20}$$

with an explicit photon propagator and the product of two currents, like

$$J_\mu = l_\mu^{\text{em}} + J_\mu^{\text{em}}. \tag{2.21}$$

The hadronic current was discussed in Chapter 1 and the leptonic current has a term for each charged lepton like

$$l_\mu^{\text{em}} = \overline{\Psi}_l \gamma_\mu \Psi_l. \tag{2.22}$$

On comparing (2.20) with (1.1), we note that the propagator is missing in (1.1). It should have been there in the form

$$g^2 \Delta_{\mu\nu} = g^2 \frac{-\mathrm{i}g_{\mu\nu}}{q^2 - M_W^2}, \tag{2.23}$$

if the weak interaction were mediated by the exchange of a particle of mass M_W and coupling strength g. At very low energies, however, at which most of the decays take place, $q^2 \ll M_W^2$ and

$$g^2 \Delta_{\mu\nu} \longrightarrow \mathrm{i} g_{\mu\nu} \frac{g^2}{M_W^2} = \mathrm{i} g_{\mu\nu} \frac{G}{\sqrt{2}}. \tag{2.24}$$

Thus the form in (1.1) is indeed a very good approximation.

2.2 The partially conserved axial current

A second property of the weak currents deals with approximations that are possible in matrix elements of the axial current. The charged pions decay weakly into $\mu\bar{\nu}_\mu$ pairs with a hadronic matrix element

$$\langle 0 | A_\mu^{\pm}(x) | \pi^{\pm}(q) \rangle = \mathrm{i} f_\pi(q^2) q_\mu \mathrm{e}^{-\mathrm{i}qx}. \tag{2.25}$$

Here q_μ is the four-momentum of the pion and f_π defines the decay coupling constant. The form of the matrix is dictated by Lorentz invariance. The coupling $f_\pi(q^2 = m_\pi^2)$ is measured with the pion on the mass shell. We can take the divergence of this matrix element and obtain

$$q^\mu \langle 0 | A_\mu^{\pm}(x) | \pi^{\pm} \rangle = \mathrm{i} f_\pi q^2 \mathrm{e}^{-\mathrm{i}qx}. \tag{2.26}$$

We conclude from this relation that the axial current is not conserved, because neither f_π nor m_π is zero. However, it may be approximately conserved because $q^2 = m_\pi^2$ is a small number relative to the mass squared of all other hadrons. Thus, for many low-energy processes that involve the axial current with four-momentum q_μ, it is possible to replace the divergence of the axial current by the pion field

$$\partial^\mu A_\mu^i = f_\pi m_\pi^2 \phi^i, \tag{2.27}$$

and, in addition, after we have extracted the pion propagator, the reduced matrix element is a slowly varying function of q^2 provided that $q^2 \lesssim m_\pi^2$. Equation (2.27) is an operator relation and holds for all matrix elements. We must be careful, however, to replace the pion field by its source $j_\pi^i = (\Box^2 + m^2)\phi^i$ and substitute for the pion–nucleon vertex the coupling

$$\langle p | j_{\pi^+} | n \rangle = \mathrm{i}\sqrt{2} g_{\pi\mathrm{NN}} \bar{u}(p') \gamma_5 u(p). \tag{2.28}$$

Several applications have established that the matrix elements of the axial current and its divergence can be treated this way. We shall describe here an application of this procedure to the matrix element in β-decay, which leads to a remarkable relation known as the Goldberger–Treiman relation. We present the derivation in

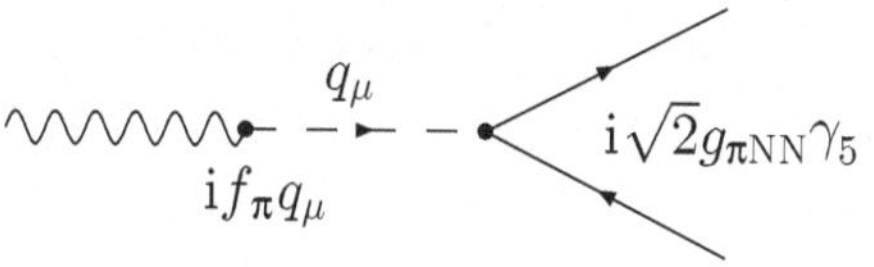

Figure 2.1. Direct coupling of the axial current to a particle of zero mass.

two ways in order to emphasize that the second method is based on an underlying symmetry. Consider the matrix element

$$\langle p|A_\mu^+|n\rangle = \bar{u}(p')\left(g_\mathrm{A}(q^2)\gamma_\mu\gamma_5 + h_\mathrm{A}(q^2)\frac{q_\mu}{2M}\gamma_5\right)u(p). \tag{2.29}$$

Taking the divergence of both sides of this equation gives

$$\langle p|\partial^\mu A_\mu^+|n\rangle = \left(2Mg_\mathrm{A}(q^2) + q^2\frac{h_\mathrm{A}(q^2)}{2M}\right)\mathrm{i}\bar{u}(p')\gamma_5 u(p). \tag{2.30}$$

On the other hand, from (2.27)

$$\begin{aligned}\langle p|\partial^\mu A_\mu^+|n\rangle &= f_\pi m_\pi^2\langle p|\phi^+|n\rangle = f_\pi m_\pi^2\frac{1}{-q^2+m_\pi^2}\langle p|j_\pi^+|n\rangle \\ &= f_\pi\frac{m_\pi^2}{-q^2+m_\pi^2}\mathrm{i}\sqrt{2}g_{\pi\mathrm{NN}}\bar{u}(p')\gamma_5 u(p).\end{aligned} \tag{2.31}$$

Taking the limit $q^2 \to 0$ with $m_\pi^2 \neq 0$ in the last two equations, we obtain

$$\sqrt{2}Mg_\mathrm{A} = g_{\pi\mathrm{NN}}f_\pi. \tag{2.32}$$

This is the Goldberger–Treiman relation. For the experimental values of the coupling constants it holds at the 10% level. It is a remarkable relation, relating the pion–nucleon coupling constant to two couplings of weak interactions.

There is a second way of looking at partially conserved axial current (PCAC). The meaning of PCAC is that the actual world is not far from the limit in which the axial currents are conserved at the expense of having zero-mass pions ($m_\pi = 0$, $f_\pi \neq 0$). In this approach we can still define f_π and g_A through Eqs. (2.25) and (2.29). Because the axial current is now conserved, Eq. (2.30) becomes

$$2Mg_\mathrm{A}(q^2) + q^2\frac{h_\mathrm{A}(q^2)}{2M} = 0. \tag{2.33}$$

In the limit of $q^2 \to 0$ the second term of Eq. (2.29) does not vanish but contributes the amplitude

$$\mathrm{i}f_\pi q_\mu\frac{\mathrm{i}}{q^2}\mathrm{i}\sqrt{2}g_{\pi\mathrm{NN}}\bar{u}(p')\gamma_5 u(p) \tag{2.34}$$

shown by the diagram in Fig. 2.1. The amplitude has a pole at $q^2 = 0$ and its divergence gives

$$q^2 \frac{h_A(q^2)}{2M} = -\sqrt{2} f_\pi g_{\pi NN} + \text{terms proportional to } q^2, \tag{2.35}$$

which, together with (2.33), gives again the Goldberger–Treiman relation. This demonstrates that the form factor $h_A(q^2)$ is dominated at small momentum transfers by the pion pole.

2.3 Regularities among the forces

The subjects covered in the first two chapters represent basic topics developed long before the electroweak theory. They strongly suggest that the weak force is not an isolated phenomenon, but one intimately connected with the other forces of nature. The isotriplet hypothesis clearly states that the isovector part of the electromagnetic current and the vector part of the weak current for $\Delta S = 0$ transitions form an isospin triplet. In addition, it states that the charges $T^{\pm}$ are the same generators as those of strong isospin. We note that operators of the three types of interactions are related. The isotriplet hypothesis also posed a problem: that of explaining why the neutral member of the multiplet did not occur in the weak interactions by itself, but only through electromagnetism. This question was answered with the discovery of weak neutral currents.

The hypothesis of PCAC relates couplings of the weak interactions to the pion–nucleon coupling constant through the Goldberger–Treiman relation. In another application, PCAC combined with equal-time commutation relations, it is possible to calculate the deviation of g_A from 1 as an integral over the pion–nucleon cross sections.

Consequences of PCAC hold at the 10%–20% level. They are understood to hold because the mass of the pion is small in comparison with the masses of other hadrons. That is, there is an underlying symmetry, which is broken by the small mass of the pion. The previous remarks provide a strong motivation to search for a closer connection of the weak, electromagnetic, and perhaps the strong interactions. The successful theory which unifies the weak and electromagnetic forces is studied in the following chapters. The electroweak theory is so far in excellent agreement with experiment. It made many predictions that have been confirmed by experimental data. Finally, the reader should keep in mind that the theory must also provide a natural explanation of the empirical rules described so far and others to be described in the following chapters.

Problems for Chapters 1 and 2

1. The scattering of particles

$$a + b \longrightarrow c + d$$

is described by the amplitude

$$f(\theta) = \frac{1}{k}\sum_l (2l+1)\frac{(\eta_l e^{2i\delta_l} - 1)}{2i} P_l(\cos\theta),$$

where η_l and δ_l are real functions and k is the magnitude of the momentum of particle a or b in the center-of-mass system. δ_l is the phase shift and η_l is introduced to describe inelastic scattering: for elastic scattering $\eta_l = 1$ and for inelastic scattering $\eta_l < 1$.

(a) Prove the optical theorem and show that

$$\sigma_{\text{tot}} = \frac{2\pi}{k^2}\sum_l (2l+1)[1 - \eta_l \cos(2\delta_l)].$$

(b) Show that, for elastic scattering,

$$\sigma_{\text{el}} = \frac{4\pi}{k^2}\sum_l (2l+1)\left|\frac{\eta_l e^{2i\delta_l} - 1}{2i}\right|^2.$$

(c) Show that, from (a) and (b), it follows that

$$\sigma_{\text{tot}} = \frac{\pi}{k^2}\sum_l (2l+1)(1 - \eta_l^2).$$

2. Using the result from Problem 1 (part (c)), show that the cross section for the reaction $\nu_\mu + e^- \to \mu^- + \nu_e$ is limited by

$$\sigma(\nu_\mu + e^- \to \mu^- + \nu_e) \le \frac{\pi}{2E_{\text{cm}}^2},$$

where E_{cm} is the energy of the ν_μ or the e^- in the center-of-mass frame. Take into account that it is an $l = 0$ scattering and that there is a spin factor of $(2s + 1)$.
3. From the Hermiticity of the electromagnetic current, show that $F_1(q^2)$ and $F_2(q^2)$ are real.
4. From time-reversal invariance, show that $F_1(q^2)$ and $F_2(q^2)$ are real.
5. By considering the non-relativistic limit of the Pauli interaction,

$$\frac{1}{2}\mu\overline{\Psi}\sigma_{\mu\nu}\Psi F^{\mu\nu}, \qquad F_{\mu\nu} = \partial_\mu A_\nu - \partial_\nu A_\mu,$$

give a physical interpretation of the term containing $F_2(q^2)$. Express $F_2(0)$ in terms of the proton's anomalous magnetic moment.

Reference

Marshak, R. E., Riazuddin, and Ryan, C. P. (1969), *Theory of Weak Interactions in Particle Physics* (New York, Interscience)

Select bibliography

For CVC and V-A interactions:

Feynman, R. P., and Gell-Mann, M. (1958), *Phys. Rev.* **109**, 1938
Gerschtein, S. S., and Zeldovich, J. R. (1955), *JETP* **29**, 698
Marshak, R. E., and Sudarshan, E. C. G. (1958), *Phys. Rev.* **109**, 1860

For PCAC:

Adler, S. L., and Dashen, R. F. (1968), *Current Algebras* (New York, W. A. Benjamin Inc.)
Cabibbo, N. (1963), *Phys. Rev. Lett.* **10**, 531
Cheng, T. P., and Li, L. F. (1984), *Gauge Theory of Elementary Particle Physics* (Oxford, Clarendon Press)
Goldberger, M. L., and Treiman, S. B. (1958), *Phys. Rev.* **110**, 1178

3

The quark model

3.1 Introduction

The quark model arose from the analysis of symmetry patterns observed when particles were grouped together according to their spin and parity. When the eight mesons with $J^p = 0^-$ are displayed in a strangeness (S) versus isospin (I_3) plane, they form the octet of Fig. 3.1. An identical pattern emerges for the eight vector mesons with $J^p = 1^-$ also shown in Fig. 3.1. The vector mesons are excited states of the particles in the $J^p = 0^-$ octet. The symmetry pattern was interpreted as a generalization of the isospin group SU(2) to the group SU(3) which incorporates both isospin and strangeness. Gell-Mann and Neeman (1964) proposed that the eight baryons with $J^p = \frac{1}{2}^+$ also belong to an octet of SU(3), thus establishing a parallelism between meson and baryon states. Finally, many static properties of the particles exhibit the SU(3) symmetry.

Since the fundamental representation of the group SU(3) is a triplet, it is natural to try to interpret the hadronic states in the octets as bound states of triplets or of triplets with antitriplets. If the fundamental fields also carry baryon number, the product of triplet $\otimes$ antitriplet would be mesons with zero baryon number. The product of three triplets carries baryon number and contain octets and a decuplet as was required by the observed states of baryons. This is the quark model of Gell-Mann (1964) and Zweig (1964).

The spectroscopy of particles and their SU(3) properties are covered in many books, for instance in the references at the end of this chapter, and we shall concentrate on symmetries of the currents, which are more relevant for the electroweak theory.

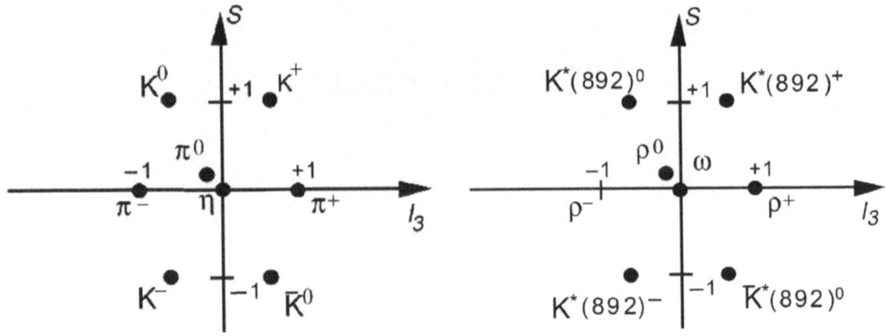

Figure 3.1. Meson octets with $J^P = 0^-$ and $J^P = 1^-$.

3.2 Current algebra

The original model contained a triplet of quarks

$$q = \begin{pmatrix} \mathrm{u} \\ \mathrm{d} \\ \mathrm{s} \end{pmatrix} \tag{3.1}$$

with the quantum numbers

Quark	Q/e	I	I_3	B	Y
u	2/3	1/2	1/2	1/3	1/3
d	−1/3	1/2	−1/2	1/3	1/3
s	−1/3	0	0	1/3	−2/3

where Q, I, I_3, B, and Y are the charge, isospin, third component of isospin, baryon number, and hypercharge, respectively. The quantum numbers of the quarks satisfy the Gell-Mann–Nishijima relation,

$$Q = T_3 + \frac{Y}{2}, \tag{3.2}$$

a rule that was established originally for hadronic states.

Next we shall rewrite the currents in terms of quark fields and formulate several of their properties. This approach is motivated by the fact that several properties of the currents and their couplings to hadrons are explained as symmetry properties of SU(3) and in many cases they are identical with predictions of the simple quark model. In fact, for a long time the quark model was used as a tool for abstracting properties and relations, whose validity is more general in field theories. In the early days the quark model was supplemented with strong interactions mediated by vector mesons in order to verify the validity of the results in theories with interactions. Among the regularities are relations between the masses of particles within a multiplet and regularities of the currents. We describe below conservation laws of the currents and outline the algebra of currents.

The electromagnetic interaction of leptons is

$$\mathcal{L}_{\text{int}}^{\text{em}} = e j_\mu^{\text{em}}(x) A^\mu(x), \tag{3.3}$$

with

$$j_\mu(x) = \bar{e}(x)\gamma_\mu e(x) + \bar{\mu}(x)\gamma_\mu \mu(x) + \cdots. \tag{3.4}$$

Similarly, we can construct the electromagnetic current of quarks,

$$\begin{aligned} j_\mu(x) &= \sum_i e_{q_i} \bar{q}_i \gamma_\mu q_i \\ &= \frac{2}{3}\bar{u}\gamma_\mu u - \frac{1}{3}\bar{d}\gamma_\mu d - \frac{1}{3}\bar{s}\gamma_\mu s \\ &= \frac{1}{2}(\bar{u}\gamma_\mu u - \bar{d}\gamma_\mu d) + \frac{1}{6}(\bar{u}\gamma_\mu u + \bar{d}\gamma_\mu d - 2\bar{s}\gamma_\mu s). \end{aligned} \tag{3.5}$$

In the last equation we separated the current into two parts, in order to show explicitly its SU(3) content. Let λ^a be the Gell-Mann matrices for SU(3), then we define vector and axial currents

$$j_\mu^a(x) = \bar{q}(x)\gamma_\mu \frac{\lambda^a}{2} q(x) \tag{3.6}$$

$$j_{\mu 5}^a(x) = \bar{q}(x)\gamma_\mu \gamma_5 \frac{\lambda^a}{2} q(x) \tag{3.7}$$

with q given in Eq. (3.1). Then the electromagnetic current in (3.5) is

$$j_\mu^{\text{em}}(x) = j_\mu^3 + \frac{1}{\sqrt{3}} j_\mu^8, \tag{3.8}$$

which reproduces Eq. (2.11).

With the quark currents it is also convenient to study the symmetries of the Lagrangian

$$\mathcal{L}_{\text{quark}} = \mathcal{L}_0 + \mathcal{L}_{\text{mass}}, \tag{3.9}$$

with

$$\mathcal{L}_0 = \mathrm{i}\bar{q}\,\partial\!\!\!/\,q \quad \text{and} \quad \mathcal{L}_{\text{mass}} = m_{\text{u}}\bar{u}u + m_{\text{d}}\bar{d}d + m_{\text{s}}\bar{s}s. \tag{3.10}$$

We now state the invariance properties under global transformations. $\mathcal{L}_0$ is invariant under the transformation

$$q \longrightarrow q' = Uq \quad \text{with} \quad U = \mathrm{e}^{\mathrm{i}\theta_\alpha \lambda^a/2}, \tag{3.11}$$

where θ_α are constants, i.e. independent of space and time. Such a transformation is called global. In proving the invariance under unitary transformations recall that $e^A e^B = e^{A+B+\frac{1}{2}[A,B]}$, provided that A, B commute with $[A, B]$. The term $\mathcal{L}_{\text{mass}}$ is not, in general, invariant under the global transformation. It becomes invariant only when all quark masses are equal:

$$m = m_{\rm u} = m_{\rm d} = m_{\rm s}. \tag{3.12}$$

A consequence of the symmetry is the conservation of all vector currents. Consider

$$\frac{\partial}{\partial x^\mu}\{\bar{u}(x, p')\gamma_\mu d(x, p)\} = \frac{\partial}{\partial x^\mu} j^\dagger_\mu(x), \tag{3.13}$$

which in momentum space becomes

$$\bar{u}(x, p')\big[\not{p}' - \not{p}\big]d(x, p) = (m_{\rm u} - m_{\rm d})\bar{u}(x, p')d(x, p). \tag{3.14}$$

This current is conserved when the two masses become equal.

Let us try to repeat this argument for axial transformations:

$$q \longrightarrow q' = Vq = e^{i\Phi_\alpha \lambda^a/2\cdot\gamma_5} q. \tag{3.15}$$

Now the kinetic term $\mathcal{L}_0$ is again invariant, but the mass term is not invariant even when the masses are equal. The Lagrangian is invariant under global γ_5 transformations when all quark masses are zero. In fact the axial current is not conserved and its divergence is

$$\frac{\partial}{\partial x^\mu} j^\dagger_{\mu 5}(x) = \bar{u}(x, p')\big[\not{p}' - \not{p}\big]\gamma_5 d(x, p) = (m_{\rm u} + m_{\rm d})\bar{u}(x, p')\gamma_5 d(x, p). \tag{3.16}$$

When the Lagrangian is invariant under the axial transformations (3.15), all quark masses must vanish and the axial current is conserved. The two cases are examples of Noether's theorem, which states that, for every continuous global transformation that leaves the Lagrangian invariant, there is a current that is conserved.

Relations of the second class are abstracted from the quark model and establish equal-time commutation relations of currents.

In quantum field theory the quark fields satisfy the following equal-time canonical anticommutation relations:

$$\begin{aligned} \{q^\dagger_\tau(x), q_{\tau'}(x')\}_{x_0=x_0'} &= \delta_{\tau\tau'}\delta^{(3)}(\vec{x} - \vec{x}'), \\ \{q_\tau(x), q_{\tau'}(x')\}_{x_0=x_0'} &= \left\{q^\dagger_\tau(x), q^\dagger_{\tau'}(x')\right\}_{x_0=x_0'} = 0, \end{aligned} \tag{3.17}$$

where τ and τ' run from 1 to 12, i.e. there are three flavors and to each of them there correspond four spinor components. One can derive equal-time commutation relations for the SU(3) currents using the identity

$$[AB, CD] = -AC\{D, B\} + A\{C, B\}D - C\{D, A\}B + \{C, A\}DB. \quad (3.18)$$

The final result, outlined in Problem 2, is

$$\left[j_\mu^a(x), j_0^b(x')\right]_{x_0=x_0'} = -f^{abc} j_\mu^c(x)\delta^{(3)}(\vec{x} - \vec{x}'). \quad (3.19)$$

On integrating this equation over three-dimensional space we arrive at

$$\left[j_\mu^a(x), Q^b(x_0)\right]_{x_0=x_0'} = \mathrm{i} f^{abc} j_\mu^c(x), \quad (3.20)$$

where $Q^b(x_0)$ is the charge corresponding to the vector current, defined by

$$Q^b(x_0) = \int \mathrm{d}^3x \, j_0^b(x). \quad (3.21)$$

It is now straightforward to derive from (3.20) the commutation relation for the charges:

$$\left[Q^a, Q^b\right] = \mathrm{i} f^{abc} Q^c. \quad (3.22)$$

Similarly, we can repeat the above steps for the axial current to obtain

$$\left[Q^a, Q_5^b\right] = \mathrm{i} f^{abc} Q_5^c, \quad (3.23)$$

$$\left[Q_5^a, Q_5^b\right] = \mathrm{i} f^{abc} Q^c. \quad (3.24)$$

We see that the vector and axial charges form an algebra that closes under commutation relations. If we define left- and right-handed charges

$$Q_{\mathrm{L,R}}^a = \frac{1}{2}(Q^a \mp Q_5^b), \quad (3.25)$$

they also satisfy the algebra

$$\left[Q_\mathrm{L}^a, Q_\mathrm{R}^b\right] = 0,$$

$$\left[Q_\mathrm{L}^a, Q_\mathrm{L}^b\right] = \mathrm{i} f^{abc} Q_\mathrm{L}^c, \quad \left[Q_\mathrm{R}^a, Q_\mathrm{R}^b\right] = \mathrm{i} f^{abc} Q_\mathrm{R}^c. \quad (3.26)$$

It says that the left-handed sector does not communicate with the right-handed sector. Thus each sector by itself forms an SU(3) algebra. The group now is $\mathrm{SU(3)_L} \times \mathrm{SU(3)_R}$, known as the chiral group. The theory based on the chiral group and the approximation that the u and d quark masses are very small, relative to those of the other quarks, is known as chiral theory. The chiral theory can explain many of the regularities observed at small masses and momenta. We shall have the opportunity to remark on the implications of such a theory in Sections 5.2 and 15.6.

We can also express the weak hadronic current given by

$$j_\mu^{\mathrm{had}} = \bar{u}\gamma_\mu(1 - \gamma_5)d \cos\theta_\mathrm{c} + \bar{u}\gamma_\mu(1 - \gamma_5)s \sin\theta_\mathrm{c} \quad (3.27)$$

in terms of the octet currents (3.6) and (3.7). Defining

$$V_\mu^a = j_\mu^a, \quad A_\mu^a = j_{\mu 5}^a, \tag{3.28}$$

one obtains

$$j_\mu^{\text{had}} = \left[\left(V_\mu^1 + \mathrm{i}V_\mu^2\right) - \left(A_\mu^1 + \mathrm{i}A_\mu^2\right)\right]\cos\theta_c + \left[\left(V_\mu^4 + \mathrm{i}V_\mu^5\right) - \left(A_\mu^4 + \mathrm{i}A_\mu^5\right)\right]\sin\theta_c \tag{3.29}$$

and we recover the $\Delta S = 0$ part of (2.17) and the $\Delta S = 1$ part of (2.18).

Expressing the hadronic currents in terms of quark fields sets them in one-to-one correspondence with the leptonic currents. The equal-time commutators give non-linear relations between observables, thus determining their relative strengths. Prominent among them are several sum rules that are valid at small and large momentum transfers.

3.3 Quantum chromodynamics

In spite of its successes, the quark model was received with a lot of skepticism because there was no experimental evidence for particles with fractional charges. To some authors this remained a mystery; to others the quarks remained a mnemonic for deriving useful rules. An additional objection concerned the fact that there was no theory describing the strong interactions among quarks. The attitude changed in the late sixties when inelastic electron–nucleon-scattering experiments provided evidence for point-like constituents, partons, within hadrons. Furthermore, correlations between electron- and neutrino-induced reactions provided evidence that the partons carried the quark quantum numbers. These topics will be studied in detail in Chapters 10 and 11. The final result was the formulation of a theory for the strong interactions whose fundamental fields are the quark and vector mesons – the gluons.

The theory of strong interactions is known as quantum chromodynamics or, in short, QCD. There are strong indications that each quark carries an additional quantum number called color; hence the name of the theory chromodynamics (color = *chroma*). The choice of names of the colors as red, white, and blue, or another triplet of names, is arbitrary but the fact that they are three in number is important. The quarks interact with each other by the exchange of vector bosons that change the colors of the quarks (Gross and Wilczek, 1973; Politzer, 1973).

We include in this section a few introductory remarks and discuss topics related to QCD in various sections of the book.

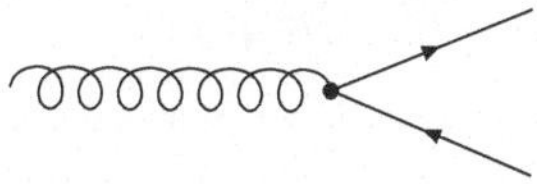

$$ig_s\gamma_\mu\frac{\lambda_a}{2}$$

Figure 3.2. A gluon–fermion vertex.

The theory of strong interactions is in many respects similar to QED. We write a quark of a specific flavor as a triplet of color SU(3):

$$q(x) = \begin{pmatrix} \mathrm{q_r} \\ \mathrm{q_w} \\ \mathrm{q_b} \end{pmatrix}.$$

The theory also contains eight vector mesons – the gluons – coupled to quarks. There is again a vector vertex, with the new coupling constant g_s for the strong interactions, and a λ^α matrix acting on the quarks. The effective coupling constant for the strong interactions,

$$\alpha_s(p) = \frac{g_s^2(p)}{4\pi},$$

is now large and calculations with the exchange of a single gluon are neither accurate nor useful. One considers the cumulative effect from the exchange of many gluons, which modify the coupling constant, making it a function of momentum carried by the gluon.

The strong coupling constant has a remarkable property. At small momenta it is very large, binding the quarks into hadrons, so quarks cannot be separated as asymptotic particles. At large momenta of the gluons, the strong coupling constant becomes small, making perturbative calculations possible. As a consequence there are two types of calculations in QCD. One of them involves large momenta, for which perturbative summations of many gluons are possible. In a second class of calculations, numerical simulations of QCD replace continuous space-time by a finite but large four-dimensional lattice for space and time. Sophisticated computer programs have been written for handling gluon and quark fields on the lattice. These are non-perturbative calculations that should produce, among other results, confinement.

Throughout this book we shall study decays and reactions that involve both strong and weak interactions. The weak interactions of hadrons will be expressed in terms of the quark substructure by writing the currents in terms of quark fields and estimating or calculating matrix elements of quark operators for transitions between hadronic states. The success of these methods varies from process to process. This is a still developing field of research, as will become evident in several sections of this book.

Problems for Chapter 3

1. The lowest-lying baryon states are built with three quarks with $L = 0$. There are ten states with $J^p = \frac{3^+}{2}$. To this decuplet belongs Δ^{++}(uuu). Construct the wave function of Δ^{++} with space, spin, and color contributions so that it obeys Fermi statistics. Finally, argue that color is necessary in order for the Pauli principle to be preserved. See Kokkedee (1969).
2. The weak vector current builds, together with the electromagnetic current, an algebra. For a better understanding we consider the SU(2) algebra. The generators of the group SU(2) are the matrices τ^1, τ^2, and τ^3, with the following property:

$$\left[\tau^i, \tau^j\right] = 2\mathrm{i}\varepsilon^{ijk}\tau_k.$$

We now define the currents

$$j_\mu^a = \bar{q}(x)\frac{\tau^a}{2}\gamma_\mu q(x),$$

which satisfy an algebra.

The fermion fields obey the canonical quantization

$$\left\{q_\tau^\dagger(x), q_{\tau'}(y)\right\}_{x_0=y_0} = \delta_{\tau\tau'}\delta^{(3)}(\vec{x}-\vec{y})$$

and

$$\{q_\tau(x), q_{\tau'}(y)\}_{x_0=y_0} = \left\{q_\tau^\dagger(x), q_{\tau'}^\dagger(y)\right\}_{x_0=y_0} = 0.$$

(a) Show the following relation:

$$\left[\Gamma_\alpha\tau^a, \Gamma_\beta\tau^b\right] = \frac{1}{2}\left\{\Gamma_\alpha, \Gamma_\beta\right\}\left[\tau^a, \tau^b\right] + \frac{1}{2}\left[\Gamma_\alpha, \Gamma_\beta\right]\left\{\tau^a, \tau^b\right\},$$

where Γ_α and Γ_β are arbitrary Dirac matrices.

(b) Using the identity

$$[AB, CD] = -AC\{B, D\} + A\{C, B\}D - C\{D, A\}B + \{C, A\}DB,$$

show that

$$\left[q_\sigma^\dagger(x)q_\tau(x), q_{\sigma'}^\dagger(y)q_{\tau'}(y)\right]_{x_0=y_0}$$
$$= \left\{q_\sigma^\dagger(x)\delta_{\sigma'\tau}q_{\tau'}(y) - q_{\sigma'}^\dagger(y)\delta_{\sigma\tau'}q_\tau(x)\right\}\delta^{(3)}(\vec{x}-\vec{y})|_{x_0=y_0}.$$

(c) It follows now that

$$\left[\mathrm{i}\bar{q}\gamma_\mu\frac{\tau^a}{2}q(x), \mathrm{i}\bar{q}\gamma_0\frac{\tau^b}{2}q(y)\right]_{x_0=y_0} = -\mathrm{i}\varepsilon^{abc}\bar{q}(x)\frac{\tau^c}{2}\gamma_\mu q(y)\delta^{(3)}(\vec{x}-\vec{y}).$$

References

Gell-Mann, M. (1964), *Phys. Lett.* **8**, 214

Gell-Mann, M., and Neeman, Y. (1964), *The Eightfold Way* (New York, Benjamin)

Gross, D. J., and Wilczek, F. (1973), *Phys. Rev. Lett.* **39**, 1343

Kokkedee, J. J. (1969), *The Quark Model* (New York, Benjamin)

Politzer, D. (1973), *Phys. Rev. Lett.* **30**, 1346
Zweig, G. (1964), CERN preprint 8182/TH-401

Select bibliography

Kokkedee, J. J. (1969), *The Quark Model* (New York, Benjamin)
Greenberg, O. W. (1964), *Phys. Rev. Lett.* **13**, 598
Han, M. Y., and Nambu, Y. (1965), *Phys. Rev.* **139B**, 1006

Part II

Field theories with global or local symmetries

4

Yang–Mills theories

4.1 The Yang–Mills field

The successful and simple theory which unifies the weak and electromagnetic interactions is based on the group SU(2) $\times$ U(1). We develop the theory in several steps. First we describe, in this chapter, the main features of a gauge theory. Then we will describe a theory containing only electrons and the corresponding neutrinos. Finally, the theory is extended to incorporate hadrons.

The structure of a Yang–Mills theory is almost completely determined by the requirement that the internal symmetry transformations of the fields can be carried out independently at different space-time points. In other words, the theory is invariant under local transformations. Let Ψ be a multiplet of n Dirac fields. The multiplets belong to representations of the group SU(N). We define a transformation of the fermion fields by

$$\begin{aligned} \Psi &\longrightarrow \Psi' = U\Psi, \\ \bar{\Psi} &\longrightarrow \bar{\Psi}' = \bar{\Psi}U^{\dagger}, \end{aligned} \tag{4.1}$$

with U a unitary matrix. We represent U by

$$U = \mathrm{e}^{\mathrm{i}\alpha_j\lambda_j/2}, \tag{4.2}$$

with $j = 1, 2, \ldots, N^2 - 1$, where λ_j are the generators of the group with $\lambda_j = \lambda_j^{\dagger}$ and the α_j are real. The generators are familiar in simple cases. When the fermions belong to the fundamental representation of SU(2), the λ_j are the Pauli matrices; for SU(3) they are the Gell-Mann matrices. We distinguish two cases:

(i) when all α_j are constant, we call it a global transformation;
(ii) when the $\alpha_j = \alpha_j(x)$ are functions of x_μ, we call it a local or gauge transformation.

The free Dirac Lagrangian

$$\mathcal{L} = \mathrm{i}\bar{\Psi}(x)\gamma_\mu\partial^\mu\Psi(x) = \mathrm{i}\bar{\Psi}(x)\not{\partial}\Psi(x) \tag{4.3}$$

is invariant under global transformations.

If we allow α_j to be a function of x, then (4.3) is no longer invariant. In fact, the Lagrangian transforms into

$$\mathcal{L} \longrightarrow \mathcal{L}' = \bar{\Psi}\mathrm{i}\gamma_\mu\left[\partial^\mu + \frac{\mathrm{i}}{2}\alpha_j^\mu(x)\lambda_j\right]\Psi, \tag{4.4}$$

with $\alpha_j^\mu(x) = \partial\alpha_j(x)/\partial x_\mu$. Following Yang and Mills, we introduce a set of vector fields $B_i^\mu(x)$ and couple them to the currents as follows:

$$\mathcal{L} = \bar{\Psi}\mathrm{i}\gamma_\mu\left[\partial^\mu + \Lambda_j B_j^\mu\right]\Psi + \mathcal{L}_B, \tag{4.5}$$

where Λ_j is a set of matrices still to be determined and $\mathcal{L}_B$ is a function of the B_j^μ terms only. Each vector field is characterized by a Lorentz index μ and an internal symmetry index j. We now demand that $\mathcal{L}$ remains invariant under the transformations (4.2) with α_j a function of x; this will require that B_j^μ transforms in such a way as to cancel out the additional term in (4.4). Let $\hat{B}_j^\mu$ be the transformed vector field. Then, for $\mathcal{L}$ to remain invariant,

$$U^+\frac{\partial}{\partial x_\mu}U + U^+\Lambda_i U B_i^\mu = \Lambda_i\hat{B}_i^\mu \tag{4.6}$$

must hold (see Problem 1). Since the λ_j terms form a complete set of $N \times N$ traceless matrices, we can attempt to write

$$\Lambda_k = \frac{\mathrm{i}}{2}e\lambda_k; \tag{4.7}$$

the imaginary i is there because the λ_k terms and the B terms are Hermitian. Considering infinitesimal transformations,

$$U \simeq 1 + \frac{\mathrm{i}}{2}\alpha_j\lambda_j, \tag{4.8}$$

with

$$\begin{aligned}\left[\frac{1}{2}\lambda_i, \frac{1}{2}\lambda_j\right] &= \mathrm{i}f_{ijk}\frac{1}{2}\lambda_k,\\ \mathrm{Tr}[\lambda_i\lambda_j] &= 2\delta_{ij},\end{aligned} \tag{4.9}$$

and f_{ijk} are the structure constants of the group. For the infinitesimal transformation we can solve for $\hat{B}_j^\mu$ in (4.6). A convenient method is to rewrite (4.6) as

$$e\lambda_i \hat{B}_i^\mu = U^+ e\lambda_i U B_i^\mu + \lambda_i \frac{\partial \alpha_i}{\partial x_\mu} \tag{4.10}$$

and then expand the unitary matrices to first order in α^i and use the relation in (4.9) to obtain

$$\hat{B}_k^\mu = B_k^\mu + f_{ijk}\alpha_i B_j^\mu + \frac{1}{e}\frac{\partial \alpha_k}{\partial x^\mu}. \tag{4.11}$$

It is convenient to introduce a covariant derivative,

$$D^\mu = \partial^\mu + \frac{\mathrm{i}}{2} e\lambda_j B_j^\mu, \tag{4.12}$$

and rewrite the fermion part in (4.5) as

$$\mathcal{L}_\mathrm{F} = \mathrm{i}\bar{\Psi} \not{D} \Psi = \mathrm{i}\bar{\Psi}\gamma_\mu \left(\partial^\mu + \frac{\mathrm{i}}{2} e\lambda_j B_j^\mu \right) \Psi. \tag{4.13}$$

Covariant derivatives are useful in generating gauge-invariant Lagrangians. A Lagrangian invariant under global transformations becomes locally gauge-invariant when all ordinary derivatives are replaced by covariant derivatives. In quantum electrodynamics this replacement is the well-known minimal-substitution law.

Next we must construct $\mathcal{L}_\mathrm{B}$. It must be Lorentz-invariant and invariant under $B \to \hat{B}$. It must also contain the kinetic term of the B_μ fields. In analogy to the procedure of obtaining gauge-invariant field strengths in electrodynamics, we define

$$F_i^{\mu\nu} = \partial^\nu B_i^\mu - \partial^\mu B_i^\nu + e f_{ijk} B_j^\mu B_k^\nu. \tag{4.14}$$

If we introduce the vector notation

$$\vec{B}^\mu = \left(B_1^\mu, B_2^\mu, \ldots, B_k^\mu \right), \tag{4.15}$$

where $k = N^2 - 1$ and

$$(\vec{A} \times \vec{B})_i = f_{ijk} A_j B_k, \tag{4.16}$$

we can write (4.11) and (4.14) as

$$\vec{F}^{\mu\nu} = \partial^\nu \vec{B}^\mu - \partial^\mu \vec{B}^\nu + e\vec{B}^\mu \times \vec{B}^\nu. \tag{4.17}$$

The last term in (4.17) does not occur in electrodynamics and is introduced to assure that $\vec{F}^{\mu\nu}$ transforms as a vector under gauge transformations. A reason for introducing a generalized $\vec{F}_{\mu\nu}$ is given in Problem 1; in the same problem we discuss the gauge invariance of $\mathcal{L}_{\mathrm{B}}$. We can now build a scalar Lagrange function for the $\vec{B}^{\mu}$ fields,

$$\mathcal{L}_{\mathrm{B}} = -\frac{1}{4}\vec{F}^{\mu\nu}\vec{F}_{\mu\nu} = -\frac{1}{4}F_i^{\mu\nu}F_{i,\mu\nu}. \tag{4.18}$$

A theory with $\mathcal{L}_{\mathrm{B}}$ alone is called a pure Yang–Mills theory.

For Lagrangians invariant under symmetries, we can also define currents of the original Lagrangian, which are given by

$$J_\alpha^\mu(x) = \frac{\partial\mathcal{L}}{\partial(\partial_\mu\Psi)}\frac{\lambda_\alpha}{2}\Psi = \mathrm{i}\bar{\Psi}(x)\gamma^\mu\frac{\lambda_\alpha}{2}\Psi(x). \tag{4.19}$$

The invariance of the theory implies that the currents are conserved. We note that these are the same currents as those we introduced in Chapter 3.

To sum up, we constructed a theory that is invariant under gauge transformations. The complete Lagrangian is

$$\mathcal{L} = \mathcal{L}_{\mathrm{F}} + \mathcal{L}_{\mathrm{B}}.$$

We found that the invariance requirements are fulfilled by introducing vector fields coupled to conserved currents.

Such a theory is a candidate for particle physics. It describes the interaction of massless fermions with massless gauge bosons. It possesses a symmetry that can be SU(2), SU(3), or a larger unitary group. The case of SU(3) is, in fact, realized in Nature as the theory of strong interactions. There the vector bosons are the gluons coupled to quarks and the symmetry is the SU(3)-color. The color symmetry remains unbroken. The electroweak theory is more complicated because it contains masses for the quarks and the gauge bosons. It is a broken symmetry, to be developed in Chapters 5–7.

4.2 Gauge invariance in scalar electrodynamics

The electrodynamic field is described by the four-vector

$$A_\mu(x) = \big(\Phi(x), \vec{A}(x)\big), \tag{4.20}$$

whose components are the standard scalar and vector potentials. The electric and magnetic fields are now determined by

$$\vec{E} = -\vec{\nabla}A_0 - \frac{\partial\vec{A}}{\partial t}, \quad \vec{B} = \vec{\nabla}\times\vec{A}. \tag{4.21}$$

However, to one set of fields $(\vec{E}, \vec{B})$ there correspond many potentials A_μ. The primed potentials obtained by the gauge transformation

$$A'_\mu(x) = A_\mu(x) + \frac{\partial \Lambda(x)}{\partial x^\mu}, \tag{4.22}$$

with $\Lambda(x)$ an arbitrary scalar function, give the same $\vec{E}$ and $\vec{B}$. If someone solves a problem with $A_\mu(x)$ and somebody else does it with $A'_\mu(x)$, both should get the same physical result. In general, only those quantities which are invariant under gauge transformations have physical meaning. Gauge invariance has far-reaching implications for the theories, as we discuss in the following chapters, and clever choices of gauge lead to substantial simplifications of problems.

Here we use gauge invariance to discuss the degrees of freedom for the electromagnetic field. For the pure electromagnetic case

$$\mathcal{L} = -\frac{1}{4} F_{\mu\nu} F^{\mu\nu}, \tag{4.23}$$

with

$$F_{\mu\nu} = \partial_\mu A_\nu - \partial_\nu A_\mu. \tag{4.24}$$

The equation of motion

$$\partial_\mu F^{\mu\nu} = 0, \tag{4.25}$$

when written in terms of A_μ, becomes

$$\partial^\mu \partial_\mu A_\nu - \partial_\nu \partial^\mu A_\mu = 0. \tag{4.26}$$

It is well known that the $\vec{E}$ and the $\vec{B}$ fields satisfy a wave equation, but A_μ does not. In order to recover a wave equation from (4.26), we impose the condition

$$\partial_\mu A^\mu = 0 \quad \text{(Lorentz gauge)}. \tag{4.27}$$

We used the gauge freedom to obtain this result, but still we did not exhaust all possible gauge transformations, because any gauge function $\chi(x)$ that satisfies

$$\partial^\mu \partial_\mu \chi(x) = 0 \tag{4.28}$$

is still consistent with (4.27). We take advantage of this freedom in order to show that a photon has only two degrees of freedom.

A free photon is represented by a plane wave

$$A_\mu(x) = \varepsilon_\mu \mathrm{e}^{-ikx}, \tag{4.29}$$

where ε_μ is called the polarization vector. By substituting (4.29) into (4.26), we find $k^2 = 0$, or $m = 0$, and from (4.27) we find

$$k_\mu \varepsilon^\mu = 0 \quad \text{(Lorentz gauge).} \tag{4.30}$$

We choose a coordinate system with the z-axis along $\vec{k}$ and decompose ε_μ into longitudinal and transverse parts:

$$A_\mu(x) = \left(\varepsilon_\mu^{\parallel} + \varepsilon_\mu^{\perp}\right) \mathrm{e}^{-ikx}. \tag{4.31}$$

From (4.30) we conclude that $\varepsilon_\mu^{\parallel}$ is proportional to k_μ. Instead of the electromagnetic field $A_\mu(x)$, we can choose another one given through a gauge transformation with

$$\chi(x) = \mathrm{i}c\mathrm{e}^{-ikx}, \qquad c = \text{constant}. \tag{4.32}$$

The new field is

$$A'_\mu(x) = \left(\varepsilon_\mu^{\parallel} + \varepsilon_\mu^{\perp}\right) \mathrm{e}^{-ikx} + ck_\mu \mathrm{e}^{-ikx}. \tag{4.33}$$

By an appropriate choice of the constant c, we can eliminate k_μ, i.e. we can gauge away the longitudinal degrees of freedom. Therefore a free photon has only two degrees of freedom.

The argument fails for a massive $A_\mu(x)$ field. The addition of a mass term $\frac{1}{2}\mu^2 A_\mu A^\mu$ to Eq. (4.23) breaks the gauge invariance of the theory. In this case the equation of motion

$$\partial_\mu F^{\mu\nu} + \mu^2 A^\nu = 0 \tag{4.34}$$

implies

$$\mu^2 \partial_\nu A^\nu = 0. \tag{4.35}$$

For $\mu^2 \neq 0$, A_μ satisfies the Lorentz condition, which again eliminates one degree of freedom. But now we cannot repeat the steps between Eqs. (4.28) and (4.33). Therefore a massive field has three degrees of freedom.

Next we study the interaction of a photon with a charged scalar field: scalar electrodynamics. The Lagrangian is

$$\mathcal{L} = -\frac{1}{4} F_{\mu\nu} F^{\mu\nu} + (D^\mu \phi)^* (D_\mu \phi) - V(\phi^* \phi). \tag{4.36}$$

Here $D^\mu = (\partial^\mu + \mathrm{i}eA^\mu)$ is the covariant derivative, in agreement with the rule of replacing ordinary derivatives with covariant ones. We represent the field as

$$\phi(x) = \frac{1}{\sqrt{2}}(\phi_1(x) + \mathrm{i}\phi_2(x)) \tag{4.37}$$

and introduce the potential

$$V(\phi^*\phi) = -\mu^2\phi^*\phi + \lambda(\phi^*\phi)^2. \tag{4.38}$$

This Lagrangian is invariant under the local transformation

$$\begin{aligned} A^\mu(x) &\longrightarrow A^\mu(x) + \partial^\mu\omega(x), \\ \phi(x) &\longrightarrow \mathrm{e}^{-\mathrm{i}\omega(x)}\phi(x), \\ \phi^*(x) &\longrightarrow \mathrm{e}^{\mathrm{i}\omega(x)}\phi^*(x), \end{aligned} \tag{4.39}$$

where $\omega(x)$ is an arbitrary real function. The photon is again massless and carries two independent degrees of freedom. This follows from the same arguments as in the free-photon case. First we can go to the Lorentz gauge, which again simplifies the equations of motion both for $A_\mu(x)$ and for the scalar field $\phi(x)$. Then, since the photon is again massless, we can introduce a new gauge transformation satisfying (4.28) and eliminate the longitudinal degrees of freedom. In gauge theories, masses for the gauge bosons are introduced, not through the ad-hoc procedure of the previous paragraph, but through spontaneous breaking of the symmetry. We study this topic in Chapter 6 and return to scalar electrodynamics in Section 5.3.

Problems for Chapter 4

1. Consider the fermion Lagrangian in Eq. (4.13).
   (i) Show that invariance under the local transformation (4.2) requires that the covariant derivative satisfies

   $$D'_\mu = U^\dagger D_\mu U.$$

   (ii) Show that this result, together with the definition of D_μ, implies the transformation property for B^μ given in (4.6).
   (iii) Show that the Hermitian quantity $F_{\mu\nu} = -\mathrm{i}[D_\mu, D_\nu]$ is the field tensor whose transformation under local transformations is

   $$F^{i'}_{\mu\nu} = U^\dagger F^i_{\mu\nu} U.$$

   It is now easy to build an invariant term given by

   $$\mathcal{L}_{\mathrm{YM}} = \frac{1}{4}\,\mathrm{Tr}\big(F^i_{\mu\nu}F^{\mu\nu}_i\big).$$

2. Show that, under local transformations, the field-strength tensor $F^i_{\mu\nu}$ transforms as a vector on the index i. The result holds including terms linear in $\varepsilon_i(x)$, i.e.

   $$\vec{F}'_{\mu\nu} = \vec{F}_{\mu\nu} - \vec{\varepsilon} \times \vec{F}_{\mu\nu} + O(\varepsilon^2).$$

You may need the Jacobi identity

$$f_{ABm}f_{mC\alpha} + f_{BCm}f_{mA\alpha} + f_{CAm}f_{mB\alpha} = 0,$$

which follows from

$$\left[\frac{\lambda_A}{2}, \frac{\lambda_B}{2}\right] = \mathrm{i} f_{ABm}\frac{\lambda_m}{2} \quad \text{and} \quad \mathrm{Tr}(\lambda_A \lambda_B) = 2\delta_{AB}.$$

Select bibliography

Anderson, P. W. (1963), *Phys. Rev.* **130**, 439
Yang, C. N., and Mills, R. (1954), *Phys. Rev.* **96**, 191

5

Spontaneous breaking of symmetries

> If my view is correct, the universe may have a kind of domain structure. In one part of the universe you may have one preferred direction of the axis; in another part the direction of the axis may be different.
>
> *(Y. Nambu)*

In the gauge theory of the previous chapter, all gauge bosons and fermions are massless. In the real world the only massless vector particle is the photon. Evidently we must devise a procedure for giving masses to gauge bosons and other particles. During the past few decades, substantial progress has been made in understanding the connection between particle masses and symmetries. In theories with global symmetries, it is possible for the states to have the same symmetry as the operators of the theory, as is, for instance, the case with strong isospin. This, however, is not the only mode in which a symmetry manifests itself. In field theories the symmetry can be broken by giving a non-vanishing vacuum expectation value to some field, i.e.

$$\langle\Omega|\phi|\Omega\rangle \neq 0.$$

We say now that the symmetry is spontaneously broken.[1] In this case the operators of the theory exhibit the symmetry, but the physical states do not. In other words, for a symmetry which is spontaneously broken, remnants of the symmetry occur explicitly in the commutation relations of the operators, but are realized in the particle spectrum in a subtle way. In this chapter we study two such cases: the Goldstone mode and the Higgs phenomenon.

In many cases a symmetry does not allow the introduction of a mass term. The breaking of the symmetry generates a mass term. This is demonstrated in Section 5.1, where the Lagrangian is invariant under a discrete symmetry. The

[1] The word spontaneous is used to communicate the idea that the phenomenon happens without any evident external cause, for example spontaneous combustion, spontaneous emission, . . .

selection of the vacuum state breaks the symmetry and at the same time generates a mass.

For theories with continuous global symmetries the situation is different. The selection of a non-trivial vacuum generates masses, but at least one of the scalar particles must remain massless. This is the Goldstone phenomenon described in Section 5.2. Finally, in gauge theories, the particles that would become Goldstone mesons are eliminated by a gauge transformation and produce masses for gauge bosons (the Higgs mechanism).

5.1 Spontaneous breaking of global symmetries: discrete symmetry

Before we describe the general case, it is instructive to discuss a few simple examples in which the main ideas are transparent. Consider a real scalar field $\phi(x)$ and the classical Lagrange function

$$\mathcal{L} = \frac{1}{2}\frac{\partial\phi}{\partial x^{\mu}}\frac{\partial\phi}{\partial x_{\mu}} - U(\phi(x)), \tag{5.1}$$

with $U(\phi)$ a potential depending on ϕ. We are interested in finding the ground state. To this end we construct the Hamiltonian

$$\mathcal{H} = \frac{1}{2}(\partial_0\phi)^2 + \frac{1}{2}(\vec{\nabla}\phi)^2 + U(\phi). \tag{5.2}$$

The field with lowest energy is a constant field, whose value minimizes the potential $U(\phi)$. All this is classical. In the quantum theory $\phi(x)$ is an operator with a conjugate momentum. The field and its conjugate momentum satisfy commutation relations. The fields operate on the eigenstates of the Hamiltonian. For simple field theories it is possible to construct the eigenstates explicitly. The lowest energy is the ground state, also called the "vacuum." The word vacuum is somewhat misleading, because the vacuum state is not empty but, rather, is a complicated superposition of many particles. The term vacuum is appropriate in free-field theory, where it corresponds to the state with no particles, but for interacting fields the vacuum is a complicated state with many particles present. The vacuum and other states for simple Hamiltonians are constructed explicitly in Problems 1–3 at the end of this chapter. In this book, by vacuum we mean the lowest energy state, which will be denoted by $|\Omega\rangle$ or simply $|\ \rangle$.

In the class of theories of Eq. (5.1) we discuss two cases:

$$U(\phi) = \frac{\lambda}{4!}\phi^4 \pm \frac{\mu^2}{2}\phi^2 \quad \text{with} \quad \lambda > 0. \tag{5.3}$$

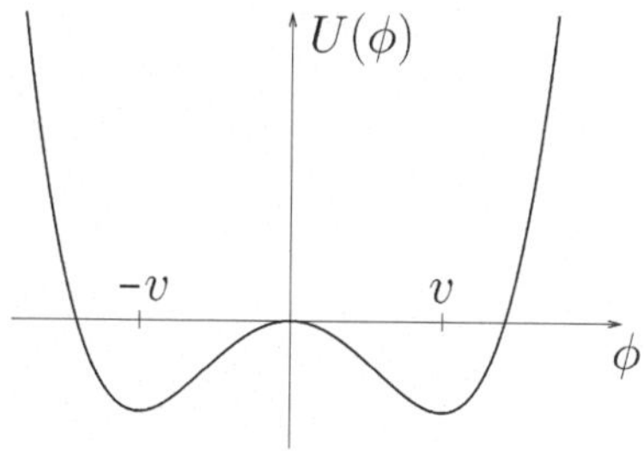

Figure 5.1. The Higgs potential $U(\phi)$.

(I) Case 1. We select

$$U(\phi) = \frac{\lambda}{4!}\phi^4 + \frac{\mu^2}{2}\phi^2. \tag{5.4}$$

This is the familiar theory for a field ϕ with mass μ^2 and an interaction term $(\lambda/4!)\phi^4$. The Feynman rules and other properties of this theory occur in many textbooks. The symmetry is explicit, with the Lagrangian being invariant under the transformation

$$\phi \to -\phi.$$

All solutions are also invariant under this transformation.

(II) Case 2. Now select

$$U(\phi) = \frac{\lambda}{4!}\phi^4 - \frac{\mu^2}{2}\phi^2. \tag{5.5}$$

In this case there is no mass term and $U(\phi)$ must be considered as a potential. The shape and the minima of the potential at

$$\phi = \pm\sqrt{\frac{3!\mu^2}{\lambda}} = \pm v$$

are shown in Fig. 5.1. We can select one of the minima as the ground state and study small oscillations around the minimum. This choice of the ground state breaks the symmetry, since the vacuum is no longer symmetric under the transformation $\phi \to -\phi$. We look for a solution in the neighborhood of v, and make the substitution

$$\phi = v + \phi' \quad \text{with} \quad v = \langle\Omega|\phi|\Omega\rangle. \tag{5.6}$$

This describes small oscillations. In terms of the new field,

$$U(\phi') = \frac{\lambda}{4!}\phi'^4 + \frac{\lambda v}{3!}\phi'^3 + \frac{1}{2}\frac{\lambda v^2}{3}\phi'^2 - \frac{\lambda v^4}{4!}.$$

We note that the term linear in ϕ' does not appear, but instead the new field acquired the mass $\sqrt{\lambda v^2/3}$. Another result of the shift is the appearance of a cubic self-coupling, which spoils the symmetry of the original Lagrangian. In this case the original symmetry is not present in the solution that we have chosen.

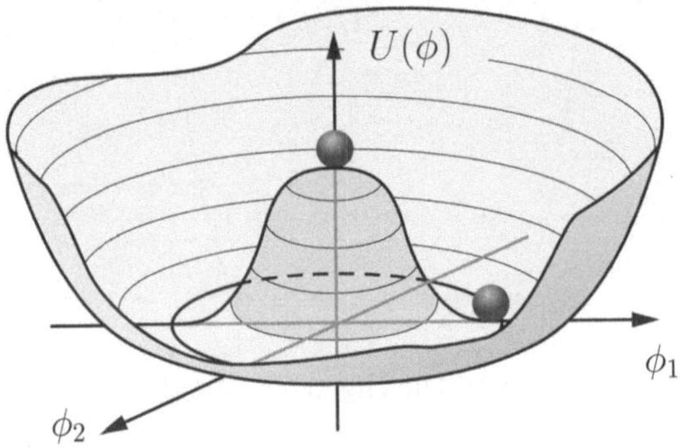

Figure 5.2. The Higgs potential $U(\phi_1, \phi_2)$.

This simple example demonstrates explicitly the breaking of the symmetry. It is based on a discreet symmetry, i.e. the reflection of the potential. New phenomena occur when the Lagrangian possesses either a continuous global symmetry or a local symmetry. In the following we discuss both cases.

5.2 Continuous global symmetries

The SO(2) model Next we consider a theory based on a continuous symmetry and then we break it spontaneously. Let us consider a theory with two real scalar fields, $\phi_1(x)$ and $\phi_2(x)$, and with the potential

$$U(\phi) = \frac{\lambda}{4!}\left(\phi_1^2 + \phi_2^2 - v^2\right)^2; \tag{5.7}$$

ϕ_1 and ϕ_2 are massless. This theory is invariant under rotation of ϕ_1 and ϕ_2, i.e. invariant under the group SO(2). The rotations are described by the angle θ,

$$\begin{pmatrix} \phi_1' \\ \phi_2' \end{pmatrix} = \begin{bmatrix} \cos\theta & \sin\theta \\ -\sin\theta & \cos\theta \end{bmatrix} \begin{pmatrix} \phi_1 \\ \phi_2 \end{pmatrix}. \tag{5.8}$$

The potential is shown in Fig. 5.2 and has the shape of a Mexican hat. The minima of the potential lie on the circle

$$\phi_1^2 + \phi_2^2 = v^2. \tag{5.9}$$

We show in Fig. 5.3 the locus of minima at the bottom of the hat. The lowest energy state is any vector $\vec{\phi}$ in the ϕ_1–ϕ_2 plane which ends at the circumference C.

We consider next the quantum-mechanical case and select a minimum in a specific direction. Without loss of generality, we select a coordinate system with the ϕ_1-axis parallel to the vacuum state, then

$$\langle\phi_1\rangle = v \quad \text{and} \quad \langle\phi_2\rangle = 0.$$

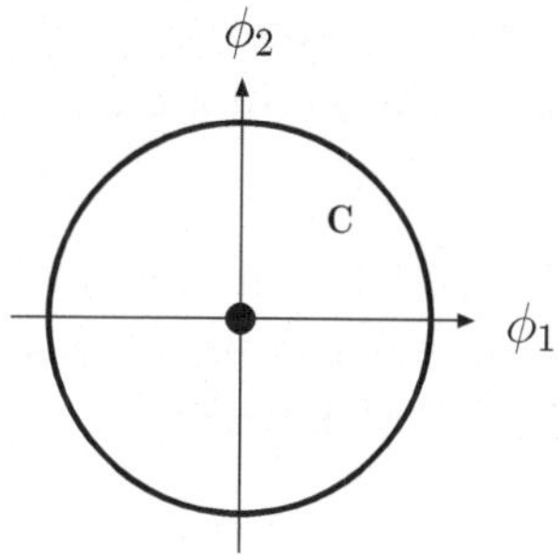

Figure 5.3. The locus of the minima of the Higgs potential.

Next we shift the fields,

$$\phi_1 = \phi_1' + v, \quad \phi_2 = \phi_2', \tag{5.10}$$

and find

$$U(\phi') = \frac{\lambda}{4!}\left(\phi_1'^2 + \phi_2'^2 + 2v\phi_1'\right)^2. \tag{5.11}$$

On expanding this, we see that the ϕ_1 field has mass, but the ϕ_2 field is massless. There again appear cubic terms in the fields, which break the original SO(2) symmetry.

This is a transparent example of a more general phenomenon and it is worthwhile to elaborate on the general case. One starts with a Lagrangian invariant under global transformations of a group G. The minima of the potential have the same symmetry. Then we break the symmetry by selecting one of the minima to be the vacuum, which is invariant under the subgroup H of G. In this analysis there are generators $\{g_i\}$ of G that do not belong to H. They are broken by the selection of the vacuum. To each *broken generator* $\{g_i\}$ there corresponds a *massless field*. These fields are called the Goldstone bosons. The Goldstone bosons transform under G like the coset or factor space of K = (G/H). We demonstrate this general phenomenon with several examples.

Example 1 In the SO(2) model, that we discussed, there is one generator

$$I(\theta) = \mathrm{e}^{\mathrm{i}\theta\sigma_y} = \begin{bmatrix} \cos\theta & \sin\theta \\ -\sin\theta & \cos\theta \end{bmatrix}. \tag{5.12}$$

After breaking of the symmetry, the potential term in (5.11) is not invariant under the transformation

$$\begin{pmatrix} \tilde{\phi}_1 \\ \tilde{\phi}_2 \end{pmatrix} = I(\theta)\begin{pmatrix} \phi_1' \\ \phi_2' \end{pmatrix}. \tag{5.13}$$

The generator is broken and there exists one massless particle ϕ_2. When we change the orientation of the vacuum, there is still a massless particle, which is a linear superposition of ϕ_1 and ϕ_2.

Example 2 We consider the model with three real fields (ϕ_1, ϕ_2, ϕ_3) invariant under the group SO(3), that is, invariant under rotations in a three-dimensional space. We represent the state by a column matrix

$$\varphi = \begin{pmatrix} \phi_1 \\ \phi_2 \\ \phi_3 \end{pmatrix} \tag{5.14}$$

and the Lagrangian by

$$\mathcal{L} = \frac{1}{2}\partial^\mu \varphi^+ \partial_\mu \varphi - U(\varphi^+\varphi), \tag{5.15}$$

with

$$U(\varphi^+\varphi) = \frac{\lambda}{4!}\left(\varphi^+\varphi - v^2\right)^2. \tag{5.16}$$

We next break the symmetry by giving a vacuum expectation value to $\langle\phi_3\rangle \neq 0$. After shifting of the fields we obtain the potential

$$U(\phi') = \frac{\lambda}{4!}\left(\phi_1'^2 + \phi_2'^2 + \phi_3'^2 + 2v\phi_3'\right)^2. \tag{5.17}$$

This expression is still invariant under rotations around the 3-axis, but $U(\phi)$ and the Lagrangian change when we rotate around the first and second axes. We can represent a general rotation through the Euler angles (α, β, γ), which consists of the following three successive rotations:

(i) a rotation through an angle α about the 3-axis with the transformation matrix

$$R(0,0,\alpha) = \begin{bmatrix} \cos\alpha & \sin\alpha & 0 \\ -\sin\alpha & \cos\alpha & 0 \\ 0 & 0 & 1 \end{bmatrix}, \tag{5.18}$$

(ii) a rotation through β about the 2-axis with $R(0, \beta, 0)$, and
(iii) a rotation through γ about the 1-axis with $R(\gamma, 0, 0)$.

The product of the three matrices gives the complete rotation

$$R(\alpha, \beta, \gamma) = R(\gamma, 0, 0)R(0, \beta, 0)R(0, 0, \alpha). \tag{5.19}$$

The model is invariant under the rotations $R(0, 0, \alpha)$, but the $R(\gamma, 0, 0)$ and $R(0, \beta, 0)$ generators are broken. To the last two generators there correspond two massless particles, as follows from the form of the potential in (5.17).

Example 3 The last example is a scalar theory invariant under global SU(2). For the field we consider a complex scalar doublet

$$\phi(x) = \begin{bmatrix} \phi_+ \\ \phi_0 \end{bmatrix} = \begin{bmatrix} \phi_1 + \mathrm{i}\phi_2 \\ \phi_3 + \mathrm{i}\phi_4 \end{bmatrix}. \tag{5.20}$$

Each of the fields has a real and an imaginary part. The Lagrangian is given by

$$\mathcal{L} = \partial_\mu \phi^\dagger \partial^\mu \phi - \lambda\left(\phi^\dagger \phi - v^2\right)^2. \tag{5.21}$$

Again we break the symmetry by giving a vacuum expectation value to the real part of ϕ_0,

$$\langle \mathrm{Re}\, \phi_0 \rangle = \langle \phi_3 \rangle = v. \tag{5.22}$$

When we shift the field ϕ_0 as before, the potential in terms of the components ϕ_+ and ϕ_0' becomes

$$U(\phi') = \lambda\left[|\phi_+|^2 + |\phi_0'|^2 + 2(\mathrm{Re}\, \phi_0')v\right]^2. \tag{5.23}$$

The new potential is not invariant under transformations of SU(2) with the two fields in $\phi_+(x)$ as well as Im ϕ_0 remaining massless, as is verified by expanding Eq. (5.23).

To sum up, we found that in field theories with continuous global symmetries the breaking of the symmetry requires the existence of scalar particles of zero mass. In fact, to every broken generator, there corresponds a massless particle. These are representative examples of Goldstone's theorem, which follows from general properties of field theory.

Goldstone's theorem If there is a continuous global symmetry transformation under which the Lagrangian is invariant, then either the vacuum state is invariant under the transformation, or there must exist spinless particles of zero mass.

We demonstrate the content of the theorem by studying the symmetry properties of a general potential.

(i) We assume that the potential $V(\phi_i)$ contains a set of real fields that transform according to a representation T^a of the group G

$$\phi_i'(x) = \phi_i(x) + \mathrm{i}\varepsilon^a T^a_{ij} \phi_j(x). \tag{5.24}$$

When the fields belong to the adjoint representation, the number of fields equals the number of generators.

(ii) We assume that the potential is invariant under the group G. Then

$$\delta V(\phi_i) = \frac{\partial V}{\partial \phi_i}\, \delta\phi_i = \mathrm{i}\frac{\partial V}{\partial \phi_i} \varepsilon^a T^a_{ij} \phi_j(x) = 0. \tag{5.25}$$

Since the ε^a are arbitrary and continuous variables, it follows that

$$\frac{\partial V}{\partial \phi_i} T_{ij}^a \phi_j(x) = 0. \tag{5.26}$$

(iii) At the minimum of the potential

$$\left.\frac{\partial V}{\partial \phi_i}\right|_{\phi_i = v_i} = 0 \tag{5.27}$$

for each ϕ_i. Differentiating Eq. (5.26) again gives

$$\frac{\partial^2 V}{\partial \phi_i \partial \phi_k} T_{ij}^a \phi_j + \frac{\partial V}{\partial \phi_i} T_{ik}^a = 0. \tag{5.28}$$

(iv) At the minimum of the potential, the second term vanishes and

$$\left.\frac{\partial^2 V}{\partial \phi_i \partial \phi_k}\right|_{\phi_i = v_i} T_{ij}^a v_j = 0. \tag{5.29}$$

The mass matrix is

$$M_{ik}^2 = \left.\frac{\partial^2 V}{\partial \phi_i \partial \phi_k}\right|_{\phi_i = v_i}. \tag{5.30}$$

Equation (5.29) is an eigenvalue equation with $T_{ij}^a v_j$ being the eigen-vectors. There are two important possibilities now. The first is

$$(\alpha) \quad T_{ij}^a v_j = 0, \tag{5.31}$$

which means that the generator T^a annihilates the vacuum. The states corresponding to the generators T^a have the symmetry of the group. In the second possibility there are generators T^b for which

$$(\beta) \quad T_{ij}^b v_j \neq 0, \tag{5.32}$$

in which case the symmetry generated by the T^b is not a symmetry of the vacuum. In this case $T_{ij}^b v_j$ is an eigenvector with eigenvalue zero, i.e. the states $T_{ij}^b \phi_j$ have zero mass. In Lagrangian theories this is a proof of the theorem, which at the same time demonstrates which particles remain massless.

Physical examples of the phenomenon occur in non-relativistic many-body systems. The Heisenberg ferromagnet is an example that consists of an infinite array of spin-$\frac{1}{2}$ magnetic dipoles. The Hamiltonian is rotationally invariant but the magnets in the ground state are aligned with all spins parallel, thus breaking the rotational symmetry. In this case the frequency of the spin waves goes to zero with the wave-number.

In particle physics the phenomenon is relevant to understanding the connection of symmetries of the hadronic currents. In Section 2.2 we showed that, in the limit of zero pion mass ($m_\pi = 0$), the axial current is conserved. The conserved vector and axial currents generate an $SU(2)_L \times SU(2)_R$ algebra. Therefore, the symmetry is present in the currents and leads to many important predictions. It is absent from the particle spectrum, since there are no parity-degenerate multiplets. Think of the (ρ^+, ρ^0, ρ^-) and (A^+, A^0, A^-) isospin multiplets. The ρs transform into each other with $SU(2)_{vector}$ and the As with $SU(2)_{axial}$, but there is no connection between the two multiplets through $SU(2)_L \times SU(2)_R$ transformations. In other words, the operators are $SU(2)_L \times SU(2)_R$-symmetric but the particle states are not ($m_\rho \neq m_A$). This physical situation can be understood in $SU(2)_L \times SU(2)_R$ theory with spontaneously broken symmetry. The $SU(2)_A$ generators are broken by non-zero vacuum expectation values and must be accompanied by zero-mass Goldstone bosons. This is in agreement with the fact that the pions have masses much smaller (nearly zero) than those of all other hadrons. Strictly speaking, the pions should be massless, but corrections to the potential or radiative corrections can produce small masses.

5.3 Spontaneous breaking of local symmetries

A new phenomenon occurs in local gauge theories, which is crucial for constructing theories with massive gauge bosons. The simplest example is scalar electrodynamics, which was introduced in Section 4.2. We consider the Lagrangian of Eq. (4.36) with the potential

$$V(\phi^*\phi) = -\mu^2\phi^*\phi + \lambda(\phi^*\phi)^2. \tag{5.33}$$

The theory is invariant under the gauge transformation defined in Eq. (4.39). In addition, the Lagrangian is invariant under a global rotation of the scalar field

$$\phi \to e^{i\alpha}\phi, \tag{5.34}$$

with α independent of space and time. We first demonstrate properties of the theory under global transformations and then indicate the changes introduced in a gauge theory.

For the vacuum state we select one of the minima of the potential. By a global rotation we can transform $\langle\phi\rangle$ to a real value. We represent the field and its vacuum state by

$$\phi = \frac{1}{\sqrt{2}}(\phi_1 + i\phi_2), \quad \langle\phi_1\rangle = v = \left(\frac{\mu^2}{2\lambda}\right)^{\frac{1}{2}} \quad \text{and} \quad \langle\phi_2\rangle = 0. \tag{5.35}$$

As in the previous cases, we translate ϕ,

$$\phi_1(x) = \phi_1'(x) + v, \tag{5.36}$$

and leave ϕ_2 unchanged. In terms of the new field, the potential becomes

$$V(\phi) = -\frac{\mu^4}{4\lambda} + \lambda\Big[\big(\phi_1'^2 + \phi_2^2\big)^2 + 4v^2\phi_1'^2 + 4v\phi_1'\big(\phi_1'^2 + \phi_2^2\big)\Big]. \tag{5.37}$$

The ϕ_1' field acquired a mass and there are also trilinear interaction terms. In addition there are changes in the kinetic terms, which we describe in Problem 5.4; important among them is the property that A_μ acquires a mass.

Alternatively, we can study this theory as a gauge theory. We define two real fields $\theta(x)$ and $\rho(x)$ by the relation

$$\phi(x) = \mathrm{e}^{\mathrm{i}\theta(x)/v}\frac{\rho(x)+v}{\sqrt{2}} \tag{5.38}$$

and give $\rho(x)$ a vacuum expectation value.

Then we observe that the local gauge transformation

$$\begin{aligned} \phi' &= \mathrm{e}^{-\mathrm{i}\theta(x)/v}\phi(x) = \frac{\rho(x)+v}{\sqrt{2}}, \\ A'_\mu(x) &= A_\mu(x) + \frac{1}{ev}\partial_\mu\theta(x) \end{aligned} \tag{5.39}$$

eliminates $\theta(x)$ completely. This transformation is unusual, because the field itself occurs in the gauge transformation; but it is legitimate in all respects. After the gauge transformation, the first two terms of Eq. (4.36) retain their form, with primed fields replacing the old ones. The net effect is

$$\mathcal{L} = -\frac{1}{4}F'_{\mu\nu}F'^{\mu\nu} + (D'^\mu\phi')^*(D'_\mu\phi') + \frac{\mu^2}{2}[\rho(x)+v]^2 - \frac{\lambda}{4}[\rho(x)+v]^4, \tag{5.40}$$

with $D'_\mu = \partial_\mu + \mathrm{i}eA'_\mu$. The second term,

$$(D'^\mu\phi')^*(D'_\mu\phi') = \frac{1}{2}\partial_\mu\rho\partial^\mu\rho + \frac{1}{2}e^2A'_\mu A'^\mu\big(\rho^2 + 2\rho v + v^2\big), \tag{5.41}$$

generates a mass for the A'_μ field. The Goldstone field disappeared and the vector field became massive.

We have described a second important case of spontaneous symmetry breaking. We started with a locally invariant theory describing a charged scalar field (two degrees of freedom) and a massless gauge field with two polarizations. After spontaneous symmetry breaking there is one real scalar field and a massive gauge field with three polarizations. The spontaneous breaking of the symmetry redistributed the degrees of freedom: one of the two real fields forming the complex scalar field was transformed into the longitudinal polarization of the vector field. This example illustrates that the spontaneous breaking of local symmetry does not produce

Goldstone mesons, but gives masses to the gauge bosons. It will be used later on in order to create masses for the intermediate gauge bosons.

The spontaneous breaking of scalar electrodynamics is physically unrealistic, because electric charge is not conserved. This is evident from the presence of the $A^2\rho$ term in (5.33). It is a consequence of the fact that we introduced a non-zero vacuum expectation value for a charged field,

$$\langle\Omega|\phi(x)|\Omega\rangle \neq 0, \tag{5.42}$$

which in itself violates charge conservation. In realistic theories we can preserve the conservation laws by giving non-zero expectation values to fields that carry the vacuum quantum numbers.

This phenomenon was introduced in the sixties in order to evade the Goldstone theorem and maintain gauge invariance, despite the fact that the vector meson acquires a mass. At that time it was thought to be relevant for the strong interactions. Later it was extended to non-Abelian gauge theories. It is now used in order to break the gauge symmetry of the electroweak theory and produce masses for the intermediate gauge bosons. It is referred to as the Higgs mechanism and we describe it in Chapter 7.

Problems for Chapter 5

1. Consider a one-dimensional harmonic oscillator. Its Hamiltonian
$$H = \frac{1}{2}(p^2 + \omega_0^2 x^2)$$
can easily be rewritten in terms of the classical variables
$$a = \sqrt{\frac{1}{2\omega_0}}(\omega_0 x + ip) \quad \text{and} \quad a^+ = \sqrt{\frac{1}{2\omega_0}}(\omega_0 x - ip).$$
In quantum mechanics a and a^+ are operators satisfying the commutation relations
$$\left[a,\, a^+\right] = 1 \quad \text{and} \quad [a,\, a] = \left[a^+,\, a\right] = 0.$$
Compute
(i) the eigenstates of this Hamiltonian,
(ii) the time development of the operators a and a^+, and
(iii) the matrix elements of a and a^+ between arbitrary states.
When you have done all this, then you have solved this quantum field theory completely.
2. The Hamiltonian for an asymmetric oscillator is
$$H = \frac{1}{2}(p^2 + \omega_0^2 x^2) + kx.$$
Replace again the position and momentum variables with the operators a and a^+. The Hamiltonian reads
$$H = \frac{1}{2}\omega_0(a^+ a + a a^+) + k(a + a^+).$$

The problem now is to find a unitary transformation such that

$$UaU^+ = a - k/\omega_0,$$
$$Ua^+U^+ = a^+ - k/\omega_0.$$

With the help of U it is possible to eliminate the linear term kx and reduce this problem to the previous one.

3. The vacuum state, $|\Omega\rangle$, is not always the empty state, $|0\rangle$. This is demonstrated with the Hamiltonian

$$H = \frac{5}{3}a^+a + \frac{2}{3}(a^+)^2 + \frac{2}{3}a^2.$$

The number operator $N = a^+a$ does not commute with the Hamiltonian. Consequently the eigenstates of N are not eigenstates of H.

Find the lowest-energy state of H. To this end, construct two operators with the properties

$$\left[H, Q^{\pm}\right] = \pm Q^{\pm}.$$

These operators raise and lower the energy by one unit. The lowest-energy state is defined as usually by the condition

$$Q^-|\Omega\rangle = 0\,.$$

You can represent the vacuum as $\sum_n c_n(a^+)^n|0\rangle$, and then use the above condition to give an explicit formula for $|\Omega\rangle$. It is possible to write the normalized vacuum state in closed form. Finally, construct all higher-energy states.

4. Work out the kinetic term for scalar electrodynamics using the new fields of Eq. (5.36). Show that the field A_μ acquired a mass and demonstrate the appearance of trilinear interaction terms.

Select bibliography

Cheng, T. P., and Li, L. F. (1992), *Gauge Theories of Elementary Particle Physics* (Oxford, Oxford University Press), see Section 8.3

Coleman, S. (1988), *Aspects of Symmetry* (Cambridge, Cambridge University Press), see Chapter 5

The original articles on the Goldstone theorem are

Goldstone, J. (1961), *Nuovo Cimento* **19**, 154

Nambu, Y. (1960), *Phys. Rev. Lett.* **4**, 380

For the Higgs mechanism, see

Higgs, P. W. (1966), *Phys. Rev.* **145**, 1156

6

Construction of the model

The next task is to find a gauge theory that contains the weak and electromagnetic currents described in the previous chapters. We consider a gauge model of electrons and their neutrinos. At the very beginning we must answer two questions:

(i) which group should we select for the theory; and
(ii) to which representation of the group should we assign the fermion fields?

The currents of the theory must include at least the charged weak current

$$J_\mu^+(x) = \bar{\nu}_e(x)\gamma_\mu(1-\gamma_5)e(x), \tag{6.1}$$

its Hermitian adjoint, and the electromagnetic current

$$J_\mu^{\mathrm{em}}(x) = \bar{e}(x)\gamma_\mu e(x). \tag{6.2}$$

We need three vector fields with which to couple them. They correspond to the intermediate gauge bosons, $W^\pm$, and the photon. The smallest group is SU(2). This group, however, is unacceptable because the currents (6.1) and (6.2) do not form an SU(2) algebra. This becomes evident on considering the charges and studying their commutation relations. Consider the charges

$$T^+ = \frac{1}{2}\int \mathrm{d}^3x\, \nu_e^+(x)(1-\gamma_5)e(x), \qquad T^- = (T^+)^+. \tag{6.3}$$

The commutator is

$$\begin{aligned}
[T^+, T^-] &= \frac{1}{4}\int \mathrm{d}^3x\, \mathrm{d}^3y \big[\nu^+(x)(1-\gamma_5)e(x), e^+(y)(1-\gamma_5)\nu(x)\big] \\
&= \frac{1}{2}\int \mathrm{d}^3x\, \mathrm{d}^3y \big[\nu^+(x)(1-\gamma_5)\nu(x) - e^+(x)(1-\gamma_5)e(x)\big]\delta^3(x-y), \\
\mathrm{i}T^3 &= [T^+, T^-],
\end{aligned} \tag{6.4}$$

which is not the charge operator corresponding to the electromagnetic current.

There are now two alternatives:

(i) introduce new leptons and modify the weak current $J_\mu^\pm$ so that we get the right SU(2) algebra, or
(ii) introduce another gauge boson W_3 and its corresponding current. In this alternative there are four gauge bosons, $W^\pm$, Z, and γ, and the group must be enlarged to $SU(2) \times U(1)$.

Both alternatives were actively studied and it became evident only after the discovery of neutral currents that Nature prefers the second solution.

We consider a theory based on the group $SU(2) \times U(1)$. We must decide how the electron and its neutrino transform under SU(2) and U(1), separately. From (6.3) and (6.4) we see that the charges

$$T^+ = \frac{1}{2}\int \mathrm{d}^3x \left[\nu^+(1-\gamma_5)e\right], \tag{6.5}$$

$$T^- = \left[T^+\right]^+, \tag{6.6}$$

$$\mathrm{i}T^3 = \frac{1}{2}\int \mathrm{d}^3x \left[\nu_\mathrm{e}^+(1-\gamma_5)\nu_\mathrm{e} - e^+(1-\gamma_5)e\right] \tag{6.7}$$

generate an SU(2) algebra. This means that the left-handed fields

$$e_\mathrm{L} = \frac{1}{2}(1-\gamma_5)e \quad \text{and} \quad \nu_\mathrm{L} = \frac{1}{2}(1-\gamma_5)\nu$$

form an SU(2) doublet,

$$\Psi_\mathrm{L} = \begin{pmatrix} \nu_\mathrm{L} \\ e_\mathrm{L} \end{pmatrix}. \tag{6.8}$$

The charges are defined as

$$T_i = \int \mathrm{d}^3x \, \Psi_\mathrm{L}^+ \left(\frac{\tau_i}{2}\right) \Psi_\mathrm{L} \tag{6.9}$$

and

$$Q = -\int \mathrm{d}^3x \, e^+ e = -\int \mathrm{d}^3x \left(e_\mathrm{L}^+ e_\mathrm{L} + e_\mathrm{R}^+ e_\mathrm{R}\right). \tag{6.10}$$

Since we should include Q in the group transformations, Q must be a combination of T_3 and the generator of U(1), denoted by Y:

$$\frac{Y}{2} \equiv Q - T_3 = -\frac{1}{2}\int \mathrm{d}^3x \left(e_\mathrm{L}^+ e_\mathrm{L} + 2e_\mathrm{R}^+ e_\mathrm{R} + \nu_\mathrm{L}^+ \nu_\mathrm{L}\right). \tag{6.11}$$

This relation involves the difference between the charge Q and the weak isospin T_3 and it defines a new quantum number: the weak hypercharge. It is analogous to the Gell-Mann–Nishijima formula of the strong interactions which was established by empirical data. With this definition the charge coincides with what we were always using as charge, but the weak isospin and hypercharge, when extended to quarks, do not always coincide with the corresponding hadronic quantum numbers. In (6.11) there appear the left-handed and right-handed leptonic number operators

$$N_{\mathrm{R}} = \int \mathrm{d}^3 x \, e_{\mathrm{R}}^{+} e_{\mathrm{R}}, \tag{6.12}$$

$$N_{\mathrm{L}} = \int \mathrm{d}^3 x \left(e_{\mathrm{L}}^{+} e_{\mathrm{L}} + \nu_{\mathrm{L}}^{+} \nu_{\mathrm{L}} \right). \tag{6.13}$$

Thus the hypercharge operator in SU(2) is the unit matrix which commutes with the other generators of the group,

$$[Y,\ T_i] = 0 \quad \text{for} \quad i = 1, 2, 3. \tag{6.14}$$

From the relation $Y = -(N_{\mathrm{L}} + 2N_{\mathrm{R}})$ we deduce that

$$Y = \begin{cases} 1 & \text{for left-handed states,} \\ 2 & \text{for right-handed states.} \end{cases}$$

This satisfies the original requirements of selecting a group and the representations for the fields with

$$\Psi_{\mathrm{L}} = \begin{pmatrix} \nu_{\mathrm{L}} \\ \mathrm{e}_{\mathrm{L}} \end{pmatrix} \qquad \text{an SU(2) doublet,}$$

$$\Psi_{\mathrm{R}} = \mathrm{e}_{\mathrm{R}} \qquad \text{an SU(2) singlet.}$$

Finally, we give the parts of the Lagrangian describing the fermion and gauge fields. We denote by W_μ^i the gauge fields of SU(2) and by B_μ the field of U(1). The field tensors are written as

$$F_{\mu\nu}^i = \partial_\mu W_\nu^i - \partial_\nu W_\mu^i + g\varepsilon^{ijk} W_\mu^j W_\nu^k, \tag{6.15}$$

$$G_{\mu\nu} = \partial_\mu B_\nu - \partial_\nu B_\mu, \tag{6.16}$$

and the Lagrangian for the gauge fields is

$$\mathcal{L}_{\mathrm{B}} = -\frac{1}{4} F_{\mu\nu}^i F^{i,\mu\nu} - \frac{1}{4} G_{\mu\nu} G^{\mu\nu}, \tag{6.17}$$

with a summation understood over repeated indices. The gauge fields at this stage are massless. The Lagrangian for the leptons is

$$\mathcal{L}_{\mathrm{F}} = \mathrm{i}\bar{\Psi}_{\mathrm{R}} \gamma_\mu \left(\partial_\mu + \mathrm{i}g' \frac{Y}{2} B_\mu \right) \Psi_{\mathrm{R}} + \mathrm{i}\bar{\Psi}_{\mathrm{L}} \gamma_\mu \left(\partial_\mu + \mathrm{i}g' \frac{Y}{2} B_\mu + \frac{\mathrm{i}}{2} g \tau^k W_\mu^k \right) \Psi_{\mathrm{L}}. \tag{6.18}$$

We notice that the leptons are also massless because there are no $\bar{\Psi}_{\mathrm{R}} \Psi_{\mathrm{L}}$ and $\bar{\Psi}_{\mathrm{L}} \Psi_{\mathrm{R}}$ terms, which indeed are not SU(2) × U(1)-invariant.

Select bibliography

Glashow, S. L. (1961), *Nucl. Phys.* **22**, 579
Salam, A., and Strathdee, J. (1972), *Nuovo Cimento* **11A**, 397
Weinberg, S. (1967), *Phys. Rev. Lett.* **19**, 1264

An early review of the electroweak theory is

Abers, E. S., and Lee, B. W. (1973), *Phys. Rep.* **9C**, 1

7

The Higgs mechanism in the Glashow–Salam–Weinberg model

7.1 Masses for gauge bosons

In order to give masses to the gauge bosons and the fermions, we follow the method described in Section 5.3. We introduce a complex scalar doublet

$$\phi = \begin{bmatrix} \phi_+ \\ \phi_0 \end{bmatrix}. \tag{7.1}$$

From the relation $Q = T_3 + Y/2$, it follows that $Y = 1$ for ϕ. Each of the fields has a real part and an imaginary part, so there are four independent scalar fields. The Lagrange function for the scalar sector is given by

$$\mathcal{L}_\phi = (D_\mu \phi)^\dagger (D^\mu \phi) - V(\phi^+ \phi), \tag{7.2}$$

with the covariant derivative defined by

$$D_\mu = \partial_\mu + \mathrm{i}g' B_\mu + \mathrm{i}g \frac{\tau^i}{2} W^i_\mu \tag{7.3}$$

and the potential by

$$V(\phi^\dagger \phi) = -\mu^2 \phi^+ \phi + \lambda (\phi^\dagger \phi)^2. \tag{7.4}$$

We have been gradually enlarging the Lagrangian and so far it consists of three terms:

$$\mathcal{L} = \mathcal{L}_\mathrm{F} + \mathcal{L}_\mathrm{B} + \mathcal{L}_\phi. \tag{7.5}$$

It contains fermions, vector bosons, and scalar fields and is invariant under gauge transformations of the group SU(2) × U(1).

Classically, the potential $V(\phi)$ has a locus of minima at

$$\frac{\partial V}{\partial \phi_+^*} = -\mu^2\phi_+ + 2\lambda\big(|\phi_+|^2 + |\phi_0|^2\big)\phi_+ = 0, \tag{7.6}$$

$$\frac{\partial V}{\partial \phi_0} = -\mu^2\phi_0 + 2\lambda\big(|\phi_+|^2 + |\phi_0|^2\big)\phi_0 = 0; \tag{7.7}$$

that is, at

$$|\phi_+|^2 + |\phi_0|^2 = \frac{\mu^2}{2\lambda}. \tag{7.8}$$

We can choose the ground state (vacuum state) at the minimum of the potential. Since we wish to conserve charge, the field must carry vacuum quantum numbers

$$|\phi_0| = \frac{\mu}{\sqrt{2\lambda}} \quad \text{and} \quad \phi_+ = 0. \tag{7.9}$$

In the quantum theory the symmetry is broken by introducing

$$\langle\phi\rangle = \begin{pmatrix} 0 \\ v/\sqrt{2} \end{pmatrix} \quad \text{with} \quad v = \frac{\mu}{\sqrt{\lambda}}.$$

In other words, one of the neutral scalar fields acquires a vacuum expectation value at the minimum of the potential.

The next step is to try to rewrite the Lagrangian in terms of fields displaced relative to the minimum of the potential and arrive at a physical interpretation. The selection of a vacuum expectation value chooses a direction in the potential, thus breaking the symmetry. We define four scalar fields, ξ_1, ξ_2, ξ_3, and η, by

$$\phi = U^{-1}(\vec{\xi})\begin{pmatrix} 0 \\ (v+\eta)/\sqrt{2} \end{pmatrix}, \tag{7.10}$$

where $U^{-1}(\vec{\xi})$ is the unitary transformation

$$U^{-1}(\vec{\xi}) = \exp\left(-\frac{\mathrm{i}\vec{\xi}\cdot\vec{\tau}}{2v}\right). \tag{7.11}$$

This is very similar to the discussion concerning Eqs. (5.38) and (5.39). Again, we define new fields through a gauge transformation

$$\phi \to \phi' = U(\vec{\xi})\phi = \begin{pmatrix} 0 \\ (v+\eta)/\sqrt{2} \end{pmatrix}, \tag{7.12}$$

$$\psi_{\mathrm{L}} \to \psi'_{\mathrm{L}} = U(\vec{\xi})\psi_{\mathrm{L}} \quad \text{and} \quad \psi'_{\mathrm{R}} = \psi_{\mathrm{R}}, \tag{7.13}$$

$$\frac{1}{2}\vec{\tau}\cdot\vec{W}_\mu \to \frac{1}{2}\vec{\tau}\cdot\vec{W}'_\mu = \frac{1}{2}U(\vec{\xi})\vec{\tau}\cdot\vec{W}_\mu U^{-1}(\vec{\xi}) + \frac{\mathrm{i}}{g}\Big[\partial_\mu U^{-1}(\vec{\xi})\Big]U(\vec{\xi}). \tag{7.14}$$

This transformation has the form described in Chapter 5, with a new feature: the fields themselves occur in the transformation. Upon substitution into the Lagrangian, the terms $\mathcal{L}_F$ and $\mathcal{L}_B$ retain the same form when expressed in terms of the new fields, but $\mathcal{L}_\phi$ is modified. In fact, as we show next, several of the scalar fields disappear and the Lagrangian has a new physical interpretation.

Consider the $\mathcal{L}_\phi$ term and set

$$\phi' = \begin{pmatrix} 0 \\ (v+\eta)/\sqrt{2} \end{pmatrix} = \frac{v+\eta}{\sqrt{2}}\chi \quad \text{with} \quad \chi = \begin{pmatrix} 0 \\ 1 \end{pmatrix}. \tag{7.15}$$

On substituting for the ϕ field in terms of the new field, it appears as if we are making a gauge transformation. The covariant derivatives become

$$D_\mu \phi' = \left[\partial_\mu \eta + \frac{\mathrm{i}}{2}(v+\eta)\big(g' B_\mu + g\tau^i W_\mu^i\big)\right]\frac{\chi}{\sqrt{2}}, \tag{7.16}$$

$$\begin{aligned}(D_\mu \phi')^\dagger (D_\mu \phi') + \text{h.c.} &= \frac{1}{2}(\partial_\mu \eta)(\partial^\mu \eta) \\ &\quad + \frac{1}{8}(v+\eta)^2 \chi^+ \big\{\big(g' B_\mu + g\tau^i W_\mu^i\big)(g' B^\mu + g\tau^i W^{i,\mu})\big\}\chi.\end{aligned} \tag{7.17}$$

The cross-term is purely imaginary and does not appear in the product. We study in detail the structure of the second term,

$$(g' B_\mu + g\vec{\tau}\cdot\vec{W}_\mu)(g' B^\mu + g\vec{\tau}\cdot\vec{W}^\mu) = \left[g'^2 B_\mu B^\mu + g^2 \vec{W}_\mu \vec{W}^\mu + 2gg' B_\mu \vec{\tau}\cdot\vec{W}^\mu\right], \tag{7.18}$$

and between the χ states

$$\begin{aligned}\chi^+[\ldots]\chi &= g'^2 B_\mu B^\mu + g^2 W_\mu^i W^{i,\mu} - 2gg' B_\mu W^{3,\mu} \\ &= \big(g' B_\mu - g W_\mu^3\big)^2 + 2g^2 W_\mu^+ W^{-\mu} \\ &= (g^2 + g'^2) Z_\mu Z^\mu + 2g^2 W_\mu^+ W^{-\mu}.\end{aligned} \tag{7.19}$$

The evaluation of the term linear in $\vec{\tau}$ is most easily done using $\vec{\tau}\cdot\vec{W}_\mu = \sqrt{2}(\tau^+ W_\mu^- + \tau^- W_\mu^+) + \tau^3 W_\mu^3$ and properties of $\tau^\pm \chi$. New fields were also introduced:

$$W_\mu^\pm = \frac{1}{\sqrt{2}}\big(W_\mu^1 \pm \mathrm{i} W_\mu^2\big), \quad Z_\mu = \frac{-g W_\mu^3 + g' B_\mu}{\sqrt{g^2 + g'^2}},$$

and

$$A_\mu = \frac{g B_\mu + g' W_\mu^3}{\sqrt{g^2 + g'^2}}. \tag{7.20}$$

We note that the fields $W^{\pm}$ and Z are now massive with

$$M_{W} = \frac{1}{2}gv \quad \text{and} \quad M_{Z} = \frac{1}{2}\left(g^2 + g'^2\right)^{1/2}v, \tag{7.21}$$

but the field A_μ remains massless. The physical correspondence for the fields is evident. A_μ represents the photon and the other three the intermediate gauge bosons of the weak interaction. An interesting property is the disappearance from the Lagrangian of the ξ_1, ξ_2, and ξ_3 fields. These three degrees of freedom were transformed into longitudinal states of the massive vector mesons. This form of the theory, with its clear physical interpretation, is referred to as the unitary gauge.

To sum up, we have constructed a theory with vector, scalar, and spin-$\frac{1}{2}$ particles based on the symmetry group SU(2) × U(1). The symmetry was broken in the Higgs mode by introducing a non-zero vacuum expectation value for the neutral field ϕ_0. Then it was shown that a judicious choice of the gauge eliminates three scalar fields. In this gauge the physical interpretation is clear, with the states $W^{\pm}$ and Z^0 being massive. They also have three longitudinal degrees of freedom.

In the quantum theory, the breaking of the symmetry by Eq. (7.9) implies that the vacuum is not the empty state but a complicated superposition of states, as demonstrated for the simple Hamiltonian in Problem 3 of Chapter 5. Condition (7.9) does not break the symmetry completely, because the charge generator annihilates the vacuum and the law for charge conservation is preserved. The other three generators are broken and to each of them there corresponds a massive gauge boson. Their masses satisfy the relation

$$\frac{M_W^2}{M_Z^2} = \frac{g^2}{g^2 + g'^2}. \tag{7.22}$$

They also couple to fermions through charged and neutral currents, which satisfy the SU(2) × U(1) algebra. These and other couplings will be studied in the following chapters.

Finally, the simple mass relations (7.21) depend on the fact that the field ϕ was a weak isodoublet. It survives even if ϕ is replaced by a finite number of isodoublet fields. It fails, however, when Higgses belonging to other representations are introduced. Consider, for instance, a theory that contains, in addition to the doublet, a triplet of Higgs fields,

$$\vec{\Sigma} = \begin{bmatrix} \Sigma^+ \\ \Sigma^0 \\ \Sigma^- \end{bmatrix}, \tag{7.23}$$

with $\langle\Sigma^0\rangle = \sigma \neq 0$. Then, by repeating the steps (7.16)–(7.21) and using the SU(2) matrices for the three-dimensional representation, the reader can verify the new mass relations

$$M_W = \frac{1}{2} g(v^2 + \sigma^2)^{1/2} \quad \text{and} \quad M_Z = \frac{1}{2}\left(g^2 + g'^2\right)^{1/2} v. \tag{7.24}$$

7.2 Masses for leptons

The standard model based on the group SU(2) × U(1) allows us to make some of the simplest choices. It is the simplest group which contains charged, neutral, and electromagnetic currents. This is at the expense of introducing two coupling constants, g and g', which are related through the Higgs mechanism to other parameters of the theory (masses of gauge bosons, structure of neutral currents, ...).

It has the simplest multiplet assignment for the fermion multiplets which is consistent with parity violation. The left-handed particles are SU(2) doublets and the right-handed components singlets,

$$\psi_e = \begin{pmatrix} \nu_e \\ e^- \end{pmatrix}_L, \quad e_R^-, \tag{7.25}$$

with the same pattern repeated for the other two families:

$$\psi_\mu = \begin{pmatrix} \nu_\mu \\ \mu^- \end{pmatrix}_L, \quad \mu_R^- \quad \text{and} \quad \psi_\tau = \begin{pmatrix} \nu_\tau \\ \tau^- \end{pmatrix}_L, \quad \tau_R^-. \tag{7.26}$$

We note that there are no right-handed neutrinos because it was thought that they are massless. This attitude changed with the discovery of neutrino oscillations, which require small and finite masses. The subject of neutrino masses is covered in Chapter 13.

Masses for leptons are generated through Yukawa couplings. A Yukawa interaction invariant under SU(2) × U(1) is given by

$$\mathcal{L}_y = g_e \bar{\psi}_e \phi e_R + \text{h.c.} \tag{7.27}$$

As mentioned earlier, the symmetry is broken by giving a vacuum expectation value to ϕ^0:

$$\begin{pmatrix} \phi^+ \\ \phi^0 \end{pmatrix} \xrightarrow{\text{breaking}} \begin{pmatrix} 0 \\ (1/\sqrt{2})(v + \eta) \end{pmatrix}, \tag{7.28}$$

which gives the mass $m_e = (1/\sqrt{2}) g_e v$. Similarly, masses are generated for the mu and tau leptons. The lepton masses remain arbitrary parameters without any relation among them.

The theory has the simplest symmetry-breaking mechanism. The Higgs particles are in the fundamental representation of SU(2) containing just enough fields to make $W^\pm$ and Z^0 massive and leave one neutral Higgs as a physical particle. This pattern of symmetry-breaking provides a consistent way to control the higher-order terms by absorbing infinities into the masses and couplings of the theory. 't Hooft (1971) derived the correct Feynman rules in a class of gauges and constructed gauges for which the theory is manifestly renormalizable. Detailed studies of renormalization and unitarity followed (Lee and Zinn-Justin, 1972; 't Hooft and Veltman, 1972). This remarkable success opened the road for many investigations and predictions that have been confirmed by experiments.

Problems for Chapter 7

1. Use the matrices

$$\lambda_1 = \frac{1}{\sqrt{2}}\begin{pmatrix} 0 & 1 & 0 \\ 1 & 0 & 1 \\ 0 & 1 & 0 \end{pmatrix}, \quad \lambda_2 = \frac{1}{\sqrt{2}}\begin{pmatrix} 0 & -\mathrm{i} & 0 \\ \mathrm{i} & 0 & -\mathrm{i} \\ 0 & \mathrm{i} & 0 \end{pmatrix} \quad \text{and}$$

$$\lambda_3 = \frac{1}{\sqrt{2}}\begin{pmatrix} 1 & 0 & 0 \\ 0 & 0 & 0 \\ 0 & 0 & -1 \end{pmatrix}$$

for the $I = 1$ representation of SU(2) and the Pauli matrices for the $I = 1/2$ representation in order to prove Eq. (7.24).

2. Generally, it is possible to construct SU(2) × U(1) theories with several multiplets of scalar fields. We denote them by ϕ_i and they carry weak isospin I_i and have a neutral component I_{3i}. If each neutral component develops a vacuum expectation value $v_i/\sqrt{2}$, show that the W and Z masses satisfy

$$M_{\mathrm{W}}^2 = \frac{1}{2} g^2 \sum_i \left[I_i(I_i + 1) - I_{3i}^2 \right] v_i^2,$$

$$M_{\mathrm{Z}}^2 = \sec^2\theta_{\mathrm{W}}\, g^2 \sum_i I_{3i}^2 v_i^2.$$

Find the first values (I, I_3) for which the relation

$$M_{\mathrm{W}} = M_{\mathrm{Z}} \cos\theta_{\mathrm{W}}$$

is maintained.

References

Lee, B. W., and Zinn-Justin, J. (1972), *Phys. Rev.* **D5**, 3121
't Hooft, G. (1971), *Nucl. Phys.* **B33**, 173; **B35**, 167
't Hooft, G., and Veltman, M. (1972), *Nucl. Phys.* **B50**, 318

Select bibliography

The following references cover the Higgs mechanism, the electroweak theory, and renormalization in the order in which they appeared.

Higgs, P. W. (1966), *Phys. Rev.* **145**, 1156
Kibble, T. W. B. (1967), *Phys. Rev.* **155**, 167
Weinberg, S. (1967), *Phys. Rev. Lett.* **19**, 1264
Salam, A., and Strathdee, J. (1972), *Nuovo Cimento* **11A**, 397
't Hooft, G. (1971), *Nucl. Phys.* **B33**, 173; **B35**, 167
Lee, B. W. (1972), *Phys. Rev.* **D5**, 3121
Lee, B. W., and Zinn-Justin, J. (1972), *Phys. Rev.* **D5**, 3121
't Hooft, G., and Veltman, M. (1972), *Nucl. Phys.* **B50**, 318

8

The leptonic sector

8.1 Feynman rules

We gave in Chapter 6 the Lagrange function for the fermions and the gauge bosons. In the previous chapter we defined the physical bosons with definite masses. It is now a straightforward exercise to rewrite the Lagrange density in terms of the physical bosons and read off the Feynman rules. For the rules, it is necessary to introduce quantized fields in order to keep track of the combinatorics and other factors, especially for diagrams with closed loops. The canonical quantization method in terms of Wick's theorem does not work for non-Abelian gauge theories because there are ambiguities that arise from gauge transformations. The appropriate discussion at this point is the quantization in the path-integral formalism. This will be a long digression and will delay us from arriving at physical results. We adopt a compromise. We consider the fermionic part of the Lagrange function in terms of the physical fields and read off the relevant vertices. The interested reader can compare this method with the procedure used in textbooks of quantum electrodynamics. In this way we obtain an extensive set of Feynman rules for vertices and propagators, in terms of which we discuss many physical processes.

Later on, we repeat this procedure for other parts of the Lagrangian, which include Higgses and gauge bosons. The rules that we obtain suffice when we calculate tree diagrams to any order. Difficulties occur when loop diagrams are computed, beginning with one-loop diagrams. The difficulties are solved by introducing additional diagrams with scalar particles: the Faddeev–Popov ghosts.

We saw in the previous chapter that the neutral gauge fields mix among themselves. It is appropriate to introduce a mixing angle

$$\tan\theta_W = \frac{g'}{g}. \tag{8.1}$$

The physical fields defined as mass eigenstates are given by

$$W_\mu^\pm = \frac{1}{\sqrt{2}}(W_1 \pm \mathrm{i}W_2), \tag{8.2}$$

$$A_\mu = \cos(\theta_\mathrm{W})B_\mu + \sin(\theta_\mathrm{W})W_\mu^3, \tag{8.3}$$

$$Z_\mu = \sin(\theta_\mathrm{W})B_\mu - \cos(\theta_\mathrm{W})W_\mu^3. \tag{8.4}$$

As stated earlier, A_μ represents the photon, while Z_μ and $W_\mu^\pm$ represent the neutral and charged intermediate gauge bosons, whose masses satisfy the relation

$$\frac{M_\mathrm{W}^2}{M_\mathrm{Z}^2} = \cos^2\theta_\mathrm{W}. \tag{8.5}$$

In the leptonic Lagrangian (6.18) we substitute for W_μ^i and B_μ in terms of the fields $W_\mu^\pm$, Z_μ, and A_μ. Using again the identity

$$\vec{\tau}\cdot\vec{W}_\mu = \sqrt{2}(\tau^+ W_\mu^- + \tau^- W_\mu^+) + \tau_3 W_\mu^3, \tag{8.6}$$

we can read off the couplings of the $\mathrm{W}^\pm$ bosons to charged currents.

The couplings of Z_μ and A_μ to their respective currents follow after some algebra. The neutral gauge couplings are

$$\begin{aligned}\mathcal{L}_\mathrm{NC} &= -\bar{\psi}_\mathrm{R}\gamma^\mu g'\frac{Y}{2}B_\mu\psi_\mathrm{R} - \bar{\psi}_\mathrm{L}\gamma^\mu\left(g'\frac{Y}{2}B_\mu + g\frac{\tau_3}{2}W_\mu^3\right)\psi_\mathrm{L} \\ &= -gs\left[\bar{\psi}_\mathrm{R}\gamma^\mu\frac{Y}{2}\psi_\mathrm{R} + \bar{\psi}_\mathrm{L}\gamma^\mu\left(\frac{Y}{2}+\frac{\tau_3}{2}\right)\psi_\mathrm{L}\right]A_\mu \\ &\quad + \frac{g}{c}\left[-s^2\left(\bar{\psi}_\mathrm{R}\gamma^\mu\frac{Y}{2}\psi_\mathrm{R} + \bar{\psi}_\mathrm{L}\gamma^\mu\frac{Y}{2}\psi_\mathrm{L}\right) + c^2\bar{\psi}_\mathrm{L}\gamma^\mu\frac{\tau_3}{2}\psi_\mathrm{L}\right]Z_\mu,\end{aligned} \tag{8.7}$$

with $(c, s) = (\cos\theta_\mathrm{W}, \sin\theta_\mathrm{W})$. The weak hypercharge Y can be replaced according to (6.11) by

$$\bar{\psi}_\mathrm{R}\frac{Y}{2}\gamma^\mu\psi_\mathrm{R} = \bar{\psi}_\mathrm{R}Q\gamma^\mu\psi_\mathrm{R}, \tag{8.8}$$

$$\bar{\psi}_\mathrm{L}\frac{Y}{2}\gamma^\mu\psi_\mathrm{L} = \bar{\psi}_\mathrm{L}\left(Q - \frac{\tau_3}{2}\right)\gamma^\mu\psi_\mathrm{L}, \tag{8.9}$$

giving finally

$$\begin{aligned}\mathcal{L}_\mathrm{NC} &= -gs(\bar{\psi}_\mathrm{R}\gamma^\mu Q\psi_\mathrm{R} + \bar{\psi}_\mathrm{L}\gamma^\mu Q\psi_\mathrm{L})A_\mu \\ &\quad + \frac{g}{c}\left[\bar{\psi}_\mathrm{L}\gamma^\mu\frac{\tau_3}{2}\psi_\mathrm{L} - s^2(\bar{\psi}_\mathrm{R}\gamma^\mu Q\psi_\mathrm{R} + \bar{\psi}_\mathrm{L}\gamma^\mu Q\psi_\mathrm{L})\right]Z_\mu.\end{aligned} \tag{8.10}$$

Finally, on substituting for

$$\psi_L = \begin{pmatrix} \nu \\ e \end{pmatrix}_L \quad \text{and} \quad \psi_R = e_R,$$

we obtain

$$\begin{aligned} \mathcal{L}_F = {} & i\bar{e}\,\not{\partial} e + i\bar{\nu}\,\not{\partial}\nu - gsA_\mu \bar{e}\gamma^\mu e \\ & + \frac{g}{2\sqrt{2}}\left[\bar{\nu}\gamma^\mu(1-\gamma_5)eW_\mu^+ + \bar{e}\gamma^\mu(1-\gamma_5)\nu W^-\right] \\ & + \frac{g}{4c}\left[\bar{\nu}\gamma^\mu(1-\gamma_5)\nu + \bar{e}\gamma^\mu\gamma_5 e - (1-4s^2)\bar{e}\gamma^\mu e\right]Z_\mu . \end{aligned} \tag{8.11}$$

The third term above is the coupling of a massless vector particle to the electromagnetic current of electrons. Its coupling is evidently the electromagnetic charge

$$e = g\sin\theta_W . \tag{8.12}$$

We can read off the following vertices:

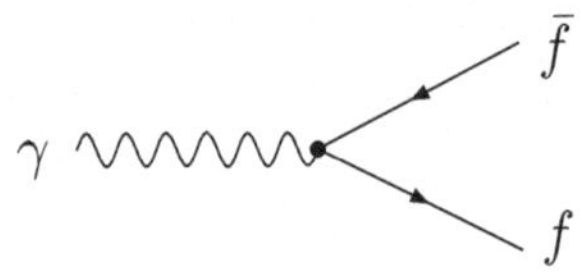

$$-ieQ_f\gamma^\mu$$

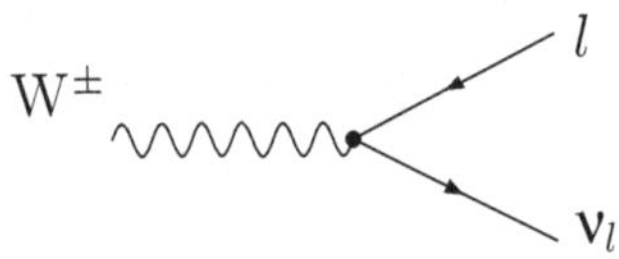

$$-\frac{ig}{2\sqrt{2}}\gamma^\mu(1-\gamma_5)$$

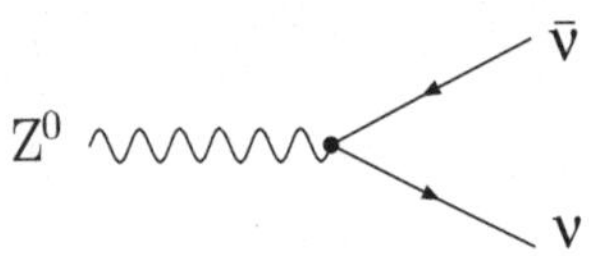

$$-\frac{ig}{4\cos\theta_W}\gamma^\mu(1-\gamma_5)$$

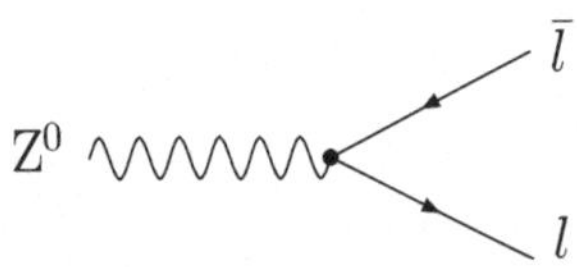

$$\frac{ig}{4\cos\theta_W}\gamma^\mu(1-4\sin^2\theta_W-\gamma_5)$$

The propagators for particles are introduced in many textbooks. For the fermions,

$$\frac{i}{\not{p} - m + i\varepsilon}$$

For vector mesons the propagator depends on the gauge. We will frequently use the Feynman gauge, for which the three relevant propagators are

γ $\quad -\dfrac{ig_{\mu\nu}}{p^2 + i\varepsilon}$

W $\quad -\dfrac{ig_{\mu\nu}}{p^2 - M_W^2 + i\varepsilon}$

Z $\quad -\dfrac{ig_{\mu\nu}}{p^2 - M_Z^2 + i\varepsilon}$

In an arbitrary gauge the vector boson propagator depends on the gauge parameter ξ as follows:

$$\Delta_{\mu\nu}(p) = -i\frac{g_{\mu\nu} + (\xi - 1)p_\mu p_\nu/(p^2 - \xi M^2)}{p^2 - M^2 + i\varepsilon}.$$

In special gauges we obtain

$$\xi = 1 \qquad \Delta_{\mu\nu} = -i\frac{g_{\mu\nu}}{p^2 - M^2 + i\varepsilon} \qquad \text{(Feynman gauge)}$$

$$\xi = 0 \qquad \Delta_{\mu\nu} = -i\frac{g_{\mu\nu} - p_\mu p_\nu/p^2}{p^2 - M^2 + i\varepsilon} \qquad \text{(Landau gauge)}$$

$$\xi = \infty \qquad \Delta_{\mu\nu} = -i\frac{g_{\mu\nu} - p_\mu p_\nu/M^2}{p^2 - M^2 + i\varepsilon} \qquad \text{(unitary gauge)}$$

In addition to electromagnetism, the theory describes weak interactions mediated by charged $W^\pm$ bosons and the neutral boson Z. Charged-current interactions are mediated by the $W^\pm$ bosons, whose mass satisfies the relation

$$M_W = \frac{1}{2}gv. \tag{8.13}$$

In low-energy reactions the W masses can be factored out – or integrated out – giving an effective four-fermion interaction with the coupling

$$\frac{G_F}{\sqrt{2}} = \frac{g^2}{8M_W^2}. \tag{8.14}$$

The electroweak theory goes beyond the V–A theory and predicts the existence of neutral currents mediated by the Z bosons, whose mass

$$M_Z = \frac{1}{2}gv\frac{1}{\cos\theta_W} \tag{8.15}$$

is related to M_{W} through (8.5). At low energies the neutral-current interaction also reduces to an effective interaction, whose overall strength is determined by

$$\left(-\frac{ig}{4\cos\theta_{\mathrm{W}}}\right)^2 \frac{1}{p^2 - M_{\mathrm{Z}}^2} = \frac{g^2}{16M_{\mathrm{W}}^2} = \frac{G_{\mathrm{F}}}{2\sqrt{2}}. \tag{8.16}$$

The Lagrangian (8.11) defines the weak and electromagnetic interactions of electrons and neutrinos. It contains four unknown quantities: g, $\sin^2\theta_{\mathrm{W}}$, and the masses M_{W} and M_{Z} which occur in the propagators of the bosons. In addition there is the mass of the electron, which is to be taken from experiment. All four quantities are not independent, since three of them are related through (8.5). We can use three experimental quantities to determine them. Electromagnetic measurements determine the fine-structure constant

$$\alpha = \frac{e^2}{4\pi} = \frac{1}{137.036\ldots}. \tag{8.17}$$

The muon decay is used to determine G_{F}. Neutral-current measurements, discussed in this and subsequent chapters, determine

$$\sin^2\theta_{\mathrm{W}} = 0.222 \pm 0.010. \tag{8.18}$$

From the three low-energy measurements we now determine the other parameters:

$$M_{\mathrm{W}}^2 = \frac{\pi\alpha}{\sqrt{2}G_{\mathrm{F}}\sin^2\theta_{\mathrm{W}}}, \quad M_{\mathrm{Z}}^2 = \frac{M_{\mathrm{W}}^2}{\cos^2\theta_{\mathrm{W}}}, \tag{8.19}$$

and

$$v = \frac{1}{\left(\sqrt{2}G_{\mathrm{F}}\right)^{1/2}} = 246\,\mathrm{GeV}. \tag{8.20}$$

8.2 Predictions in the leptonic sector

We have now a theory that enables us to compute many processes at the tree level. In this chapter we compute three leptonic processes.

Boson decays Among the many predictions of the model, the decays of the gauge bosons are simple to discuss. We begin with the decay

$$\mathrm{W}^- \to \mathrm{e}^- \bar{\nu}. \tag{8.21}$$

The diagram in Fig. 8.1 gives the amplitude

$$\mathcal{M} = \frac{ig}{2\sqrt{2}}\bar{u}(k_-)\gamma_\mu(1-\gamma_5)v(k_+)\varepsilon^\mu, \tag{8.22}$$

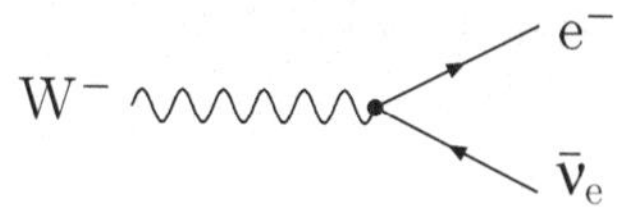

Figure 8.1. The Feynman diagram for a W^- decay.

with ε^μ being the polarization of the intermediate boson. For the decay rate we sum over the polarization states with

$$\sum_{\text{Pol}} \varepsilon^*_\mu \varepsilon_\nu = -g_{\mu\nu} + \frac{q_\mu q_\nu}{M_W^2}. \tag{8.23}$$

The contribution from $q_\mu q_\nu$ is proportional to the masses of the leptons and can be neglected. We also ignore terms proportional to lepton masses in the trace computation. The mass of the neutrino being small or zero does not cause any difficulties, because in the spinor normalization $\bar{u}u = 2m$ they simply do not occur in the formulas. Had we used another normalization, i.e. $\bar{u}u = 1$, then we could give a small mass to the neutrinos and proceed to calculate decay rates and cross sections, but we will find in the end that the neutrino masses drop out of the formulas.

The square of the matrix element summed over spins is

$$\begin{aligned}\sum_{\text{Spins, Pol}} \mathcal{M}\mathcal{M}^* &= \frac{g^2}{8} \cdot 2 \cdot \mathrm{Tr}[\gamma_\mu(1-\gamma_5)(\not{k}_+ + m_e)\gamma_\nu(\not{k}_- - m_\nu)] \cdot (-g^{\mu\nu}) \\ &= 2g^2 k_+ \cdot k_-. \end{aligned} \tag{8.24}$$

The decay rate is given following standard rules:

$$\begin{aligned}\Gamma &= \frac{1}{2M_W}\int g^2 k_+ \cdot k_-(2\pi)^4\delta^4(p - k_+ - k_-)\frac{1}{2E_e}\frac{d^3k_-}{(2\pi)^3}\frac{1}{2E_\nu}\frac{d^3k_+}{(2\pi)^3} \\ &= \frac{g^2}{16\pi}M_W = \frac{G_F M_W^3}{2\sqrt{2}\pi}. \end{aligned} \tag{8.25}$$

We must still average over the initial polarizations of the gauge bosons and obtain

$$\overline{\Gamma} = \frac{1}{(2s+1)}\Gamma = \frac{G_F M_W^3}{6\sqrt{2}\pi}, \tag{8.26}$$

which gives 211.3 MeV for $M_W = 80$ GeV (235.9 MeV for $M_W = 83$ GeV).

The total width is obtained by adding the additional decays into $\mu\bar{\nu}_\mu$, $\tau\bar{\nu}_\tau$, and quark pairs. We introduce the vertices of the gauge bosons to quarks in the next chapter, but we mention here that the decay of W^- into a quark pair of definite color is also given by (8.26). Thus, for three generations of quarks and leptons the total width is obtained by multiplying the width by 12: (3 lepton families) + [(3 quark families) × (3 colors)] = 12;

$$\Gamma_{\text{total}} = \frac{2G_F M_W^3}{\sqrt{2}\pi}. \tag{8.27}$$

In leptonic decays all that can be observed is the charged lepton. In hadronic decays the W bosons are inferred by reconstructing the hadronic jets, which imitate to some extent the kinematic characteristics of the original quarks.

In contrast, the leptonic decays of Z bosons into charged leptons are identified by the invariant mass of the pairs. By comparing the coupling constants we obtain the partial decay widths

$$\Gamma(Z \to \nu\bar{\nu}) = \frac{G_F M_Z^3}{12\sqrt{2}\pi} \tag{8.28}$$

and

$$\Gamma(Z \to e^+e^-) = \frac{G_F M_Z^3}{12\sqrt{2}\pi}\left(1 - 4\sin^2\theta_W + 8\sin^4\theta_W\right). \tag{8.29}$$

For the total width, we need in addition the decay width into quarks (see Problem 1). Summing again over three generations,

$$\Gamma_{\text{total}}(Z) = \frac{G_F M_Z^3}{3\sqrt{2}\pi}\left(\frac{21}{4} - 10\sin^2\theta_W + \frac{40}{3}\sin^4\theta_W\right). \tag{8.30}$$

It is worth noting that the total width is sensitive to the total number of quarks and leptons lighter than M_Z and precise measurements of Γ_{total} could produce exotic surprises. The width of the Z boson has been determined in the CERN experiments to be

$$\Gamma_Z = (2.490 \pm 0.007)\,\text{GeV}.$$

The width in turn limits the number of neutrinos to

$$N_\nu = 3.09 \pm 0.13,$$

which is very close to the number of neutrinos allowed by nucleosynthesis arguments, $N_\nu \simeq 3$–4.

8.3 Leptonic neutral currents

A striking piece of evidence for the electroweak theory was the discovery of neutral currents. The Lagrangian in (8.11) describes both charged- and neutral-current interactions of neutrinos and electrons. All the couplings depend on the SU(2) coupling constant g and the Weinberg angle θ_W. At low energies, the overall strength of the neutral-current interaction is determined by G_F through Eq. (8.16). Thus all neutral-current interactions must depend on a single parameter, $\sin^2\theta_W$. There is a large number of neutral-current reactions that have been measured and the agreement after two and a half decades of research is indeed impressive. This section describes in detail leptonic neutral-current reactions. The reader will find this section very useful also for the semileptonic interactions discussed in Chapters 10–12, since many of the formulas can be taken over. We consider first neutrino–electron scattering. Six reactions of this type are shown in Fig. 8.2.

Reactions (1) and (2) can proceed only through neutral currents. Other reactions like

$$\nu_e(k) + e^-(p) \to \nu_e(k') + e^-(p') \tag{8.31}$$

involve both charged- and neutral-current diagrams. For low-energy reactions it is convenient to write the Feynman amplitude in the form

$$\mathcal{M} = -i\frac{G_F}{\sqrt{2}}\left[\bar{\nu}\gamma_\mu(1-\varepsilon\gamma_5)\nu\bar{e}\gamma^\mu(g_V - g_A\gamma_5)e\right]. \tag{8.32}$$

In this form both vertices retain the charge of the lepton. Evidently, not all reactions are of this form, because several of them include also charged-current reactions. Charged- and neutral-current reactions have different propagators and in addition the order of the spinors is different. Charged-current reactions can be transformed into the charge-retaining form by Fierz's reordering theorem. A special form of the theorem states that, when at least one of the couplings is $(1 \pm \gamma_5)$, then we can interchange the first and the third (or second and fourth) spinors. All examples in Fig. 8.2 involve a neutrino or antineutrino whose vertex is $\gamma_\mu(1-\varepsilon\gamma_5)$ with $\varepsilon = 1$ for neutrinos and $\varepsilon = -1$ for antineutrinos. As an illustration, consider the reaction (8.31) to which the two diagrams in Fig. 8.3 contribute. The amplitude is

$$\begin{aligned}
\mathcal{M} &= \left(\frac{ig}{4c}\right)^2 \frac{-i}{q^2 - M_Z^2 + i\varepsilon}\bar{u}(k')\gamma_\mu(1-\gamma_5)u(k)\cdot\bar{u}(p')\left[\gamma^\mu\gamma_5 - (1-4s^2)\gamma^\mu\right]u(p) \\
&\quad + \left(\frac{ig}{2\sqrt{2}}\right)^2 \frac{-i}{q^2 - M_W^2 + i\varepsilon}\bar{u}(p')\gamma_\mu(1-\gamma_5)u(k)\cdot\bar{u}(k')\gamma^\mu(1-\gamma_5)u(p) \\
&\xrightarrow{q^2 \ll M_W^2} -i\frac{G_F}{\sqrt{2}}\bar{u}(k')\gamma_\mu(1-\gamma_5)u(k)\cdot\bar{u}(p')\left[\left(\frac{1}{2}+2s^2\right)\gamma^\mu - \frac{1}{2}\gamma^\mu\gamma_5\right]u(p) \\
&= -i\frac{G_F}{\sqrt{2}}\bar{u}(k')\gamma_\mu(1-\gamma_5)u(k)\cdot\bar{u}(p')\left(g_V\gamma^\mu - g_A\gamma^\mu\gamma_5\right)u(p).
\end{aligned} \tag{8.33}$$

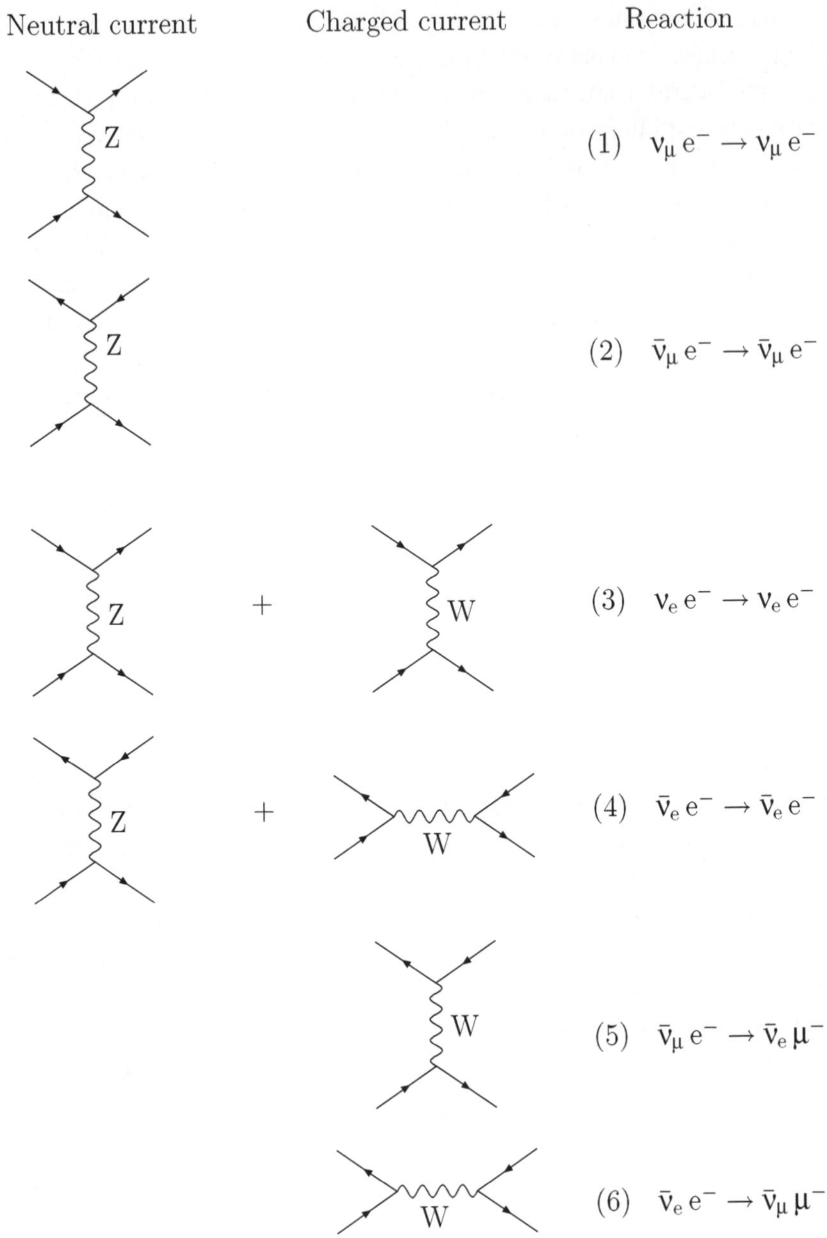

Figure 8.2. Diagrams for neutrino–electron scattering.

In this way we can write all reactions in the form (8.32). The explicit values for ε, g_V, and g_A for five reactions are given in Table 8.1. We can now calculate the cross section for the amplitude (8.32) and obtain the specific reaction by substituting the values from Table 8.1.

Table 8.1. *Effective couplings for several reactions*

Reaction	ε	Electroweak theory g_V	Electroweak theory g_A	V–A theory g_V	V–A theory g_A
$\nu_\mu + e^- \to \nu_\mu + e^-$	+1	$-\frac{1}{2} + 2s^2$	$-\frac{1}{2}$	0	0
$\bar{\nu}_\mu + e^- \to \bar{\nu}_\mu + e^-$	−1	$-\frac{1}{2} + 2s^2$	$-\frac{1}{2}$	0	0
$\nu_e + e^- \to \nu_e + e^-$	+1	$+\frac{1}{2} + 2s^2$	$+\frac{1}{2}$	1	1
$\bar{\nu}_e + e^- \to \bar{\nu}_e + e^-$	−1	$+\frac{1}{2} + 2s^2$	$+\frac{1}{2}$	1	1
$\nu_\mu + e^- \to \mu^- + \nu_e$	+1	1	1	1	1

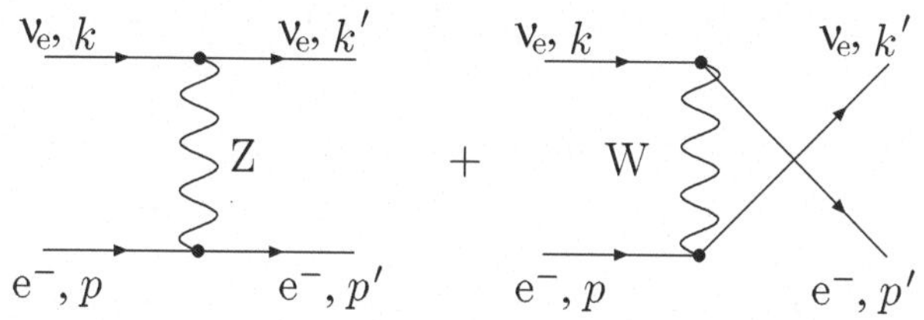

Figure 8.3. Z and W exchange in neutrino–electron scattering.

We begin with the amplitude in (8.32) and compute the differential cross section. For simplicity, for the moment we set $\varepsilon = 1$ and the mass of the electron to zero (whenever allowed). At the end we give the complete formula with ε and a small term proportional to the electron mass. The square of the amplitude summed over final spins and averaged over initial spins is

$$\begin{aligned}
|\overline{\mathcal{M}}^2| &= \frac{1}{2}\sum_{\text{Spins}} \mathcal{M}\mathcal{M}^* \\
&= \frac{G_F^2}{4}\mathrm{Tr}\big[\gamma_\mu(1-\gamma_5)\not{k}\gamma_\nu(1-\gamma_5)\not{k}'\big] \\
&\quad \times \mathrm{Tr}\big[\gamma^\mu(g_V - g_A\gamma_5)(\not{p}+m)\gamma^\nu(g_V - g_A\gamma_5)(\not{p}'+m)\big]. \quad (8.34)
\end{aligned}$$

Averaging over initial spins brings in a factor of 1/2 because the neutrinos are always left-handed. Evidently the expression factorizes into two tensors,

$$|\overline{\mathcal{M}}^2| = \frac{G_F^2}{2}\mathcal{L}_{\mu\nu}\ell^{\mu\nu}, \quad (8.35)$$

with

$$\begin{aligned}
\mathcal{L}_{\mu\nu} &= \mathrm{Tr}\big[\gamma_\mu(1-\gamma_5)\not{k}\gamma_\nu\not{k}'\big] \\
&= 4(k_\mu k'_\nu + k_\nu k'_\mu - k\cdot k' g_{\mu\nu} + \mathrm{i}\varepsilon_{\mu\nu\alpha\beta}k^\alpha k'^\beta) \quad (8.36)
\end{aligned}$$

and

$$\ell_{\mu\nu} = \mathrm{Tr}\left[(g_V^2 + g_A^2)\gamma^\mu \not{p} \gamma^\nu \not{p}' + 2g_V g_A \gamma_5 \gamma^\mu \not{p} \gamma^\nu \not{p}'\right], \tag{8.37}$$

where we neglected terms proportional to m_e^2. The computation of the $\ell_{\mu\nu}$ tensor is similar to that of $\mathcal{L}_{\mu\nu}$. In contracting the two tensors, we observe that products with different symmetries in μ and ν vanish:

$$|\overline{\mathcal{M}}|^2 = 16G_F^2\left[(g_V + g_A)^2(k \cdot p)(k' \cdot p') + (g_V - g_A)^2(k \cdot p')(k' \cdot p)\right]. \tag{8.38}$$

We choose to compute the cross section in the laboratory frame, where the initial electron is at rest:

$$d\sigma = \frac{1}{2m_e}\frac{1}{2E_\nu}|\overline{\mathcal{M}}|^2(2\pi)^4\delta^4(k + p - k' - p')\frac{1}{2E'_\nu}\frac{d^3k'}{(2\pi)^3}\frac{1}{2E'_e}\frac{d^3p'}{(2\pi)^3}. \tag{8.39}$$

We perform the d^3p' phase-space integration with the help of the δ^4-function. The last integration over the scattering angle involves the integral

$$\int d^3k'\,\delta\left[m_e^2 - (k - k' + p)^2\right] = \pi\left(\frac{E'_\nu}{E_\nu}\right)dE'_\nu. \tag{8.40}$$

The δ-function gives the relation between the scattering angle and the energy transfer:

$$1 - \cos\theta = \frac{(E - E')m_e}{EE'}. \tag{8.41}$$

Since the average value for $\langle E'\rangle \approx E_\nu/2$, we can estimate the average scattering angle $\theta \approx 2°/\sqrt{E}$ with E measured in GeV. The scattered electron comes out at very small forward angles and provides a unique signature for the experiments.

From (8.38) and (8.40) we obtain the final result

$$\frac{d\sigma}{dE'} = \frac{G_F^2 m_e}{2\pi}\left[(g_V + g_A)^2 + (g_V - g_A)^2\left(\frac{E'_\nu}{E_\nu}\right)^2\right]. \tag{8.42}$$

Had we used the amplitude (8.32) and retained the mass of the electron, the final result would have been

$$\frac{d\sigma}{dE'} = \frac{G_F^2 m_e}{2\pi}\left[(g_V + \varepsilon g_A)^2 + (g_V - \varepsilon g_A)^2\left(\frac{E'_\nu}{E_\nu}\right)^2 + \frac{m_e\nu}{E_\nu^2}(g_A^2 - g_V^2)\right]. \tag{8.43}$$

The last term, which is proportional to the electron mass, is small, so it can be neglected at accelerator energies. The variable $\nu = E_\nu - E'_\nu$ denotes the energy transfer.

Formula (8.43) is useful in describing all reactions shown in Fig. 8.2 that include charged-current reactions. For instance the reaction

$$\nu_\mu + e^- \to \mu^- + \nu_e \tag{8.44}$$

reduces after substitution to

$$\frac{d\sigma}{dE'} = \frac{2G_F^2 m_e}{\pi}. \tag{8.45}$$

The same result holds for the semileptonic reaction

$$\nu_\mu + d \to \mu^- + u, \tag{8.46}$$

which we rewrite in terms of the inelasticity $y = (E - E')/E$ and the square of the center-of-mass energy, $s = 2ME_\nu$, as

$$\frac{d\sigma^\nu}{dy} = \frac{G_F^2 s}{\pi}. \tag{8.47}$$

For both of the reactions

$$\bar{\nu}_\mu + u \to \mu^+ + d, \tag{8.48}$$

$$\nu_\mu + \bar{u} \to \mu^- + \bar{d} \tag{8.49}$$

the differential cross section is

$$\frac{d\sigma^{\bar{\nu}}}{dy} = \frac{G_F^2 s}{\pi}(1 - y)^2. \tag{8.50}$$

We see that, when both vertices are left-handed or both are right-handed, then $d\sigma/dy$ is independent of y as in (8.47). On the other hand, when one vertex is left-handed and the other right-handed, then $d\sigma/dy$ is proportional to $(1 - y)^2$ as in (8.50).

Neutral-current reactions have a mixed y dependence. As an illustrative example, consider the neutral-current reaction

$$\nu_\mu + e^- \to \nu_\mu + e^-, \tag{8.51}$$

for which the cross section is

$$\frac{d\sigma}{dy} = \frac{G_F^2 m_e E_\nu}{2\pi}\left[(1 - 2\sin^2\theta_W)^2 + 4\sin\theta_W(1 - y)^2\right]. \tag{8.52}$$

This reaction has been studied in several experiments. When the existence of neutral currents was still in doubt, a few events of this type were observed in the Gargamelle experiment. In spite of many attempts, one could not attribute them to any other origin. This evidence was gradually reinforced by information from semileptonic neutral-current reactions until their existence was accepted. Today there are experimental results from a few hundred events for reaction (8.51). The average slope from all the experiments is

$$\frac{\sigma}{E_\nu} = (1.6 \pm 0.4) \times 10^{-42}\,\mathrm{cm}^2\,\mathrm{GeV}^{-1},$$

yielding a Weinberg angle given by

$$\sin^2\theta_\mathrm{W} = 0.222 \pm 0.010.$$

Data for the other neutrino reactions are also available. In the g_V versus g_A plane each of the above total cross sections limits the physical region to an elliptical band.

8.4 Weak effects in electron–positron annihilation

An interesting and very important reaction occurs in electron–positron collisions,

$$\mathrm{e}^+\mathrm{e}^- \to \mu^+\mu^-,$$

which can be mediated by the exchange of a photon, as well as by the heavy boson Z^0. At low energies relative to the mass of the Z boson the photon diagram dominates. On the other hand, at a center-of-mass energy close to the Z boson's mass, the gauge boson dominates and exhibits a resonance behavior. At intermediate energies there is an interference term between electromagnetic and weak interactions, which modifies the angular distribution.

The amplitude for the process has two diagrams producing two amplitudes:

$$\begin{aligned}
\mathcal{M} &= \mathcal{M}_\gamma + \mathcal{M}_\mathrm{Z};\\
\mathcal{M}_\gamma &= -\frac{e^2}{s}\,\bar{v}(k_+)\gamma_\mu u(k_-)\,\bar{v}(p_+)\gamma^\mu u(p_-),\\
\mathcal{M}_\mathrm{Z} &= -\frac{g^2}{4\cos^2\theta_\mathrm{W}}\frac{1}{q^2 - M_\mathrm{Z}^2 + \mathrm{i}M_\mathrm{Z}\Gamma}\\
&\quad \times \bar{v}(k_+)\gamma_\mu(g_\mathrm{V} + g_\mathrm{A}\gamma_5)u(k_-)\bar{v}(p_+)\gamma^\mu(g'_\mathrm{V} + g'_\mathrm{A}\gamma_5)u(p_-).
\end{aligned} \tag{8.53}$$

We consider energies high enough that one can ignore the masses of the electron and muon. We also included the width of the Z particle in the propagator, which will lead to a cross section with a Breit–Wigner formula. In field theory the width is generated by summing the decays to all possible final states. For the couplings of the Z boson to electrons we introduced general coupling constants g_V and g_A; their dependence on the Weinberg angle follows from the Feynman rules. Similarly, g'_V and g'_A are couplings to muons, which are equal to the couplings of electrons.

At intermediate energies, such as $\sqrt{s} = 30$–$50\,\mathrm{GeV}$, the well-known electromagnetic formula is modified by the presence of the interference term

$$\frac{\mathrm{d}\sigma}{\mathrm{d}\Omega} = \frac{\alpha^2}{4s}\left[(1+\cos^2\theta)\left(1+\varepsilon(s)g_\mathrm{V}^2\right) + 2\,\varepsilon(s)g_\mathrm{A}^2\cos\theta\right], \tag{8.54}$$

with s the square of the center-of-mass energy and

$$\varepsilon(s) = \frac{\sqrt{2}Gs}{4\pi\alpha}. \tag{8.55}$$

The new feature is the $\cos\theta$ term, with the consequence that the differential cross section is not symmetric in the forward–backward direction. The new term arises from the neutral current but is not parity–violating. It has a new angular dependence typical of the γ_5 coupling. We define a forward–backward asymmetry

$$A(\theta) = \frac{\mathrm{d}\sigma(\theta) - \mathrm{d}\sigma(\pi-\theta)}{\mathrm{d}\sigma(\theta) + \mathrm{d}\sigma(\pi-\theta)} = \varepsilon(s)\frac{2\cos\theta}{1+\cos^2\theta}g_\mathrm{A}^2\,. \tag{8.56}$$

The asymmetry was measured in many experiments and gave values for $g_\mathrm{A} = -\frac{1}{2}$ consistent with the standard model.

At higher energies the interference term becomes larger. As the center-of-mass energy approaches the mass

$$M_\mathrm{Z} = 91.188 \pm 0.002\,\mathrm{GeV}/c^2,$$

the weak term dominates and produces the cross section

$$\frac{\mathrm{d}\sigma}{\mathrm{d}\Omega} = \left(\frac{g^2}{8c^2}\right)^2 \frac{s}{\left(s-M_\mathrm{Z}^2\right)^2 + M_\mathrm{Z}^2\Gamma^2}\left[\left(g_\mathrm{V}^2+g_\mathrm{A}^2\right)^2(1+\cos^2\theta) + 8g_\mathrm{V}^2g_\mathrm{A}^2\cos\theta\right], \tag{8.57}$$

where $c = \cos\theta_\mathrm{W}$. The resonance was observed at CERN and was studied carefully to give precise values for the mass and the width of the gauge boson quoted in this chapter.

In addition to the muons, electron–positron collisions also produce $\mathrm{q}\bar{\mathrm{q}}$ pairs, which are analyzed with the same formulas. The values for g_V' and g_A' are now replaced by couplings appropriate for quarks.

Problem for Chapter 8

1. Compute for each generation the decay width of the Z boson for decays to neutrinos, charged leptons, and quark pairs separately. Then estimate the total decay width.

9
Incorporating hadrons

9.1 The mixing matrix

The weak interaction for the leptons was introduced into the theory by arranging the left-handed leptons (chirality −1) in three generations of doublets and the right-handed charged leptons into three singlets:

$$\begin{pmatrix} \nu_e \\ e^- \end{pmatrix}_L, \quad \begin{pmatrix} \nu_\mu \\ \mu^- \end{pmatrix}_L, \quad \begin{pmatrix} \nu_\tau \\ \tau^- \end{pmatrix}_L; \quad e_R^-, \ \mu_R^-, \ \tau_R^-. \tag{9.1}$$

The similarities of the interactions of leptons to those of quarks suggest that one should similarly introduce for the quarks left-handed doublets and right-handed singlets. The situation for the quarks is different, since all of them are massive. For this reason each quark field has a right-handed component. The fields are classified as three doublets,

$$q_L^{1'} = \begin{pmatrix} u' \\ d' \end{pmatrix}_L, \quad q_L^{2'} = \begin{pmatrix} c' \\ s' \end{pmatrix}_L, \quad \text{and} \quad q_L^{3'} = \begin{pmatrix} t' \\ b' \end{pmatrix}_L; \tag{9.2}$$

and six right-handed singlets, u_R', d_R', c_R', s_R', t_R', and b_R'. The superscripts denote three generations and the primes indicate that they are gauge quarks. The part of the Lagrangian which contains the kinetic terms and the couplings of the quarks to $W^\pm$, Z^0, and photons is written as follows:

$$\mathcal{L} = \bar{q}'_{Li}\gamma^\mu\left\{i\partial_\mu + \frac{g}{2}\vec{\tau}\vec{W}_\mu + \frac{g'}{2}YB_\mu\right\}q'_{Li} + \bar{q}'_{Ri}\gamma^\mu\left\{i\partial_\mu + \frac{g'}{2}YB_\mu\right\}q'_{Ri}. \tag{9.3}$$

The operator Y denotes the weak hypercharge, which has been defined already,

$$Y = 2(Q - I_3). \tag{9.4}$$

At this stage it is not clear whether the fields u', d', ... stand for the physical states because Eq. (9.3) contains only kinetic and interaction terms. Physical fields are eigenstates of the mass matrix which will be introduced below. This is the reason why the quark fields in Eq. (9.3) have a prime and we referred to them as gauge eigenstates or gauge quarks.

Masses for the quarks are generated through quark–Higgs Yukawa couplings. A Yukawa interaction invariant under SU(2) ⊗ U(1) gauge transformations is easily constructed:

$$\mathcal{L}_{\text{mass}} = h^{ij}_{(\text{d})}(\bar{u}'_i\,,\ \bar{d}'_i)_{\text{L}}\begin{pmatrix}\phi^+\\ \phi^0\end{pmatrix} d_{j\text{R}} + h^{ij}_{(\text{u})}(\bar{u}'_i\,,\ \bar{d}'_i)\begin{pmatrix}-\bar{\phi}^0\\ \phi^-\end{pmatrix} u_{j\text{R}} + \text{h.c.}, \tag{9.5}$$

where $i,\ j = 1,\ 2,\ 3$ and the Higgs fields are the same fields as those introduced in Chapters 5 and 7. The matrices $h^{ij}_{(\text{u})}$ and $h^{ij}_{(\text{d})}$ denote couplings of i and j quarks of the up and down types, respectively. The symmetry is broken by giving a vacuum expectation value to ϕ^0:

$$\phi = \begin{pmatrix}\phi^+\\ \phi^0\end{pmatrix} \underset{\text{breaking}}{\longrightarrow} \phi = \frac{1}{\sqrt{2}}\begin{pmatrix}0\\ v+\eta\end{pmatrix} \tag{9.6}$$

and

$$\phi_{\text{c}} = -\text{i}\tau_2\phi^* = \begin{pmatrix}-\bar{\phi}^0\\ \phi^-\end{pmatrix} \underset{\text{breaking}}{\longrightarrow} -\frac{1}{\sqrt{2}}\begin{pmatrix}v+\eta\\ 0\end{pmatrix}, \tag{9.7}$$

with η the field fluctuation around the minimum. Spontaneous breaking of the symmetry generates the mass terms

$$\mathcal{L}_{\text{mass}} = \frac{v}{\sqrt{2}}\left(\bar{u}'_{\text{L}i}h^{(\text{u})}_{ij}u'_{\text{R}j} + \bar{d}'_{\text{R}j} + \bar{d}'_{\text{L}i}h^{(\text{d})}_{ij}d'_{\text{R}j}\right) + \text{h.c.} \tag{9.8}$$

The expressions

$$M^{(\text{u})}_{ij} = \frac{v}{\sqrt{2}}h^{(\text{u})}_{ij} \qquad \text{and} \qquad M^{(\text{d})}_{ij} = \frac{v}{\sqrt{2}}h^{(\text{d})}_{ij} \tag{9.9}$$

occurring above are called the mass matrices. From the way they were introduced, there is no reason for them to be either symmetric or Hermitian. In fact the Lagrangian in (9.5) is manifestly gauge-invariant and the mass matrices are to a certain extent arbitrary. The mass matrices are very important because they determine the masses and the flavor mixing of the quarks.

The quark fields that have been investigated up to now are, as has already been mentioned, non-physical gauge eigenstates. To find the physical or mass eigenstates, we must transform the quark-mass matrices into diagonal form.

Any square matrix can be diagonalized by a bi-unitary transformation. Therefore it is always possible to find four matrices $\text{U}_{\text{L,R}}$ and $\text{D}_{\text{L,R}}$ that diagonalize the mass matrices,

$$\mathcal{M}^{(\mathrm{u})} = U_{\mathrm{L}}^{+} M^{(\mathrm{u})} U_{\mathrm{R}} = \begin{pmatrix} m_{\mathrm{u}} & 0 & 0 \\ 0 & m_{\mathrm{c}} & 0 \\ 0 & 0 & m_{\mathrm{t}} \end{pmatrix}, \tag{9.10}$$

$$\mathcal{M}^{(\mathrm{u})} = D_{\mathrm{L}}^{+} M^{(\mathrm{d})} D_{\mathrm{R}} = \begin{pmatrix} m_{\mathrm{d}} & 0 & 0 \\ 0 & m_{\mathrm{s}} & 0 \\ 0 & 0 & m_{\mathrm{b}} \end{pmatrix}. \tag{9.11}$$

The mass eigenstates, to be denoted as unprimed fields, are related to the gauge eigenstates by the transformations

$$\begin{aligned} u_{\mathrm{L}i} &= (U_{\mathrm{L}}^{+})_{ij} u'_{\mathrm{L}j}, \quad u_{\mathrm{R}i} = (U_{\mathrm{R}}^{+})_{ij} u'_{\mathrm{R}j}, \\ d_{\mathrm{L}i} &= (D_{\mathrm{L}}^{+})_{ij} d'_{\mathrm{L}j}, \quad d_{\mathrm{R}i} = (D_{\mathrm{R}}^{+})_{ij} d'_{\mathrm{R}j}. \end{aligned} \tag{9.12}$$

In terms of the mass eigenstates the mass term is now diagonal. Therefore we can substitute the physical fields everywhere in the Lagrangian and deduce the physical couplings. The neutral couplings expressed in terms of physical quarks retain the same form as they had with gauge quarks. The charge current after the substitution becomes

$$j_{\mu}^{+} = \bar{u}'_{\mathrm{L}i} \gamma_{\mu} d'_{\mathrm{L}i} = \bar{u}_{\mathrm{L}i} \gamma_{\mu} V_{ij} d_{\mathrm{L}j} \tag{9.13}$$

and the charged-current interaction is

$$\mathcal{L}_{\mathrm{cc}} = \frac{g}{\sqrt{2}} (j_{\mu}^{+} W_{\mu}^{-} + j_{\mu}^{-} W_{\mu}^{+}), \tag{9.14}$$

where $V = U_{\mathrm{L}}^{+} D_{\mathrm{L}}$ and summation over repeated indices is understood. This matrix is one of the most important quantities in the standard model, because it contains information on all possible flavor-transitions and CP violation. It is called the flavor-mixing matrix or the Cabibbo (1963)–Kobayashi–Maskawa (1973) (CKM) matrix.

By construction the flavor matrix is unitary, a property that will be used extensively in the next section. The mixing matrix is derived directly from the mass matrices, which shows that all information about it is included in the mass matrices. A determination of the mass matrices from experimental data is impossible because they contain 36 real parameters (9 complex numbers for each charge sector). By contrast, there are only ten quantities that can be determined by experiment: six quark masses and four independent mixing parameters. The fact that only four parameters of the mixing matrix are relevant can be understood as follows: a unitary $N \times N$ matrix may be expressed by N^2 real parameters. Among them $N(N-1)/2$ can be chosen to be the rotation angles of an orthogonal matrix and the remaining

$N(N+1)/2$ taken as phase angles. Not all phases, however, are physical. Each quark field has an arbitrary phase that can be used to eliminate a phase of the CKM matrix, except for an overall phase. This means that one can arrange the phases of the quark fields in such a way that they eliminate phases in V_{ij} of Eq. (9.14). An exception to this rule is an overall phase that is lost when we square matrix elements in order to produce probabilities. Thus $N \times N$ flavor mixing can be parametrized by $N(N-1)/2$ rotation angles and

$$N(N+1)/2 - (2N-1) = (N-1)(N-2)/2$$

phase angles. For three quark generations ($N = 3$) there remain three rotation angles and one phase, which is responsible for CP violation. So in all there are four parameters that describe the mixing matrix.

The unitarity of the mixing matrix that is required in gauge theories is a consequence of the unitarity of the matrices U_{L} and D_{L}:

$$V_{ik}(V_{kj})^{+} = V_{ik}V_{jk}^{*} = \delta_{ij}. \tag{9.15}$$

This relation expresses the orthogonality of rows and columns within the matrix.

As was shown above, the mixing matrix can be parametrized with four quantities, but this does not determine the functional form of the matrix. The first explicit parametrization was given by Kobayashi and Maskawa. They used Euler-type angles for three-dimensional rotations in the flavor space and one phase:

$$V_{\mathrm{KM}} = \begin{pmatrix} c_1 & -s_1c_3 & -s_1s_3 \\ s_1c_2 & c_1c_2c_3 - s_2s_3\mathrm{e}^{\mathrm{i}\delta} & c_1c_2s_3 + s_2c_3\mathrm{e}^{\mathrm{i}\delta} \\ s_1s_2 & c_1s_2c_3 + c_2s_3\mathrm{e}^{\mathrm{i}\delta} & c_1s_2s_3 - c_2c_3\mathrm{e}^{\mathrm{i}\delta} \end{pmatrix}, \tag{9.16}$$

where the abbreviations $s_i = \sin\theta_i$ and $c_i = \cos\theta_i$ are used. The parameters are chosen so that, for $\theta_2 = \theta_3 = \delta = 0$, the three-dimensional mixing matrix is reduced to the corresponding one for just two doublets. The angles θ_i may without loss of generality be chosen to lie in the first quadrant, $0 \leq \theta_i < \pi/2$. The phase angle δ may take any value within the interval $[-\pi, \pi]$.

This parametrization is just one possibility. Another one that is very customary was given by Maiani. It is quite suitable for investigations of B-meson decays:

$$V_{\mathrm{M}} = \begin{pmatrix} c_\beta c_\theta & c_\beta s_\theta & s_\beta \\ -s_\beta s_\gamma c_\theta \mathrm{e}^{\mathrm{i}\delta'} - s_\theta c_\gamma & c_\gamma c_\theta - s_\beta s_\gamma s_\theta \mathrm{e}^{\mathrm{i}\delta'} & s_\gamma c_\beta \mathrm{e}^{\mathrm{i}\delta'} \\ -s_\beta c_\gamma c_\theta + s_\gamma s_\theta \mathrm{e}^{-\mathrm{i}\delta'} & -s_\beta s_\theta c_\gamma - s_\gamma c_\theta \mathrm{e}^{-\mathrm{i}\delta'} & c_\beta c_\gamma \end{pmatrix}. \tag{9.17}$$

The quantity s_γ is mainly the coupling for $\mathrm{b} \rightarrow \mathrm{c}$ and s_β is mainly for $\mathrm{b} \rightarrow \mathrm{u}$. Ranges for angles and the phase can be chosen as in the Kobayashi–Maskawa parametrization.

A further parametrization is the one by Wolfenstein. In this case the elements are expanded in terms of a small parameter $\lambda = \sin\theta_c$ exploiting the experimental information about the smallness of the mixing angles. The structure of the matrix is determined by the unitary conditions of the mixing matrix:

$$V_W = \begin{pmatrix} 1-\frac{1}{2}\lambda^2 & \lambda & A\lambda^3(\rho - i\eta + i\eta\frac{1}{2}\lambda^2) \\ -\lambda & 1-\frac{1}{2}\lambda^2 - i\eta A^2\lambda^4 & A\lambda^2(1+i\eta\lambda^2) \\ A\lambda^3(1-\rho-i\eta) & -A\lambda^2 & 1 \end{pmatrix}. \tag{9.18}$$

In contrast to the parametrizations discussed so far, the one by Wolfenstein is only an approximation, being an expansion in a small parameter. The unitarity condition is satisfied to a given order in λ. The real parts of the elements are correct to the order λ^3 and the imaginary parts to order λ^5. The parameters A, η, and ρ are of order unity or even smaller:

$$\lambda = 0.221 \pm 0.002, \quad A = 1.0 \pm 0.1, \quad \sqrt{\rho^2+\eta^2} = 0.46 \pm 0.23. \tag{9.19}$$

Magnitudes for elements of the mixing matrix are determined directly from experiments. They in turn are translated into values for the rotation angles. We will review the experiments and the corresponding values. It is much more difficult to obtain values for δ, since it is related to CP-violating quantities. We shall return to its determination in Chapters 15 and 16.

9.2 Flavor-changing neutral couplings (FCNCs)

The structure of the mixing matrix has another important consequence. The neutral couplings of the theory preserve flavor to a large degree of accuracy. The suppression of flavor-changing neutral couplings is required by many experimental results. For instance, the decay $K_L \to \mu^+\mu^-$ is highly suppressed. The branching ratio is

$$\mathrm{Br}(K_L \to \mu^+\mu^-) = (7.2 \pm 0.2) \times 10^{-9}.$$

The $D^0\bar{D}^0$ mixing has not been observed at the 10^{-3} level. These properties and others are incorporated into the theory by the construction described in Section 9.1. In fact, FCNCs are absent at the tree level. The proof of this follows from the structure of neutral currents. We can begin again with the Lagrangian in Eq. (8.10) and substitute the ψs with quark fields. The couplings of the quarks to the Z and γ are

$$\mathcal{L}_{nc} = -e\sum_{i=1}^{3} \bar{q}^{\,i} Q\gamma^\mu q_i A_\mu + \frac{g}{c}\sum_{l=1}^{3}\{\bar{q}^{\,i}\tau_3\gamma^\mu q_i - s^2\bar{q}^i Q\gamma^\mu q_i\} Z_\mu. \tag{9.20}$$

Table 9.1. *Couplings of quarks and leptons to* Z^0

States	g_V	g_A
Up quarks	$\frac{1}{2} - \frac{4}{3}\sin^2\theta_W$	$\frac{1}{2}$
Down quarks	$-\frac{1}{2} + \frac{2}{3}\sin^2\theta_W$	$-\frac{1}{2}$
Neutrinos	$\frac{1}{2}$	$\frac{1}{2}$
Charged leptons	$-\frac{1}{2} + 2\sin^2\theta_W$	$-\frac{1}{2}$

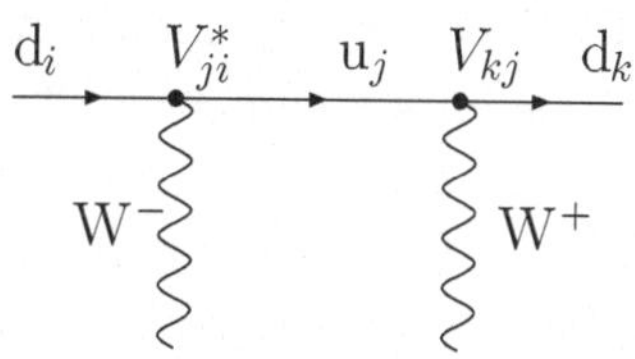

Figure 9.1. Vertices appearing in box diagrams.

At first sight the quarks occurring in (9.20) should have a prime, since they are still gauge quarks, but Eq. (9.20) is diagonal in the quark fields and the unitary matrices $U_{L,R}$ and $V_{L,R}$ will disappear when the quark fields are replaced by physical states. Thus the omission of the prime is justified.

Next we introduce a convenient notation and write the neutral-current couplings in the form

$$\mathcal{L}_{nc} = \frac{g}{2\sqrt{2}\,c}\bar{q}^i\gamma^\mu(g_V - g_A\gamma_5)q^i Z_\mu, \tag{9.21}$$

with q^i representing general states of up and down quarks or leptons. The couplings are given in Table 9.1, where we also include the neutral couplings to neutrinos and leptons. In this way neutral couplings are diagonal at the tree level.

The suppression that has been introduced so far is not sufficient. Flavor-changing effects will now appear through higher-order corrections that involve charged currents. Higher-order effects are $O(G\alpha)$ and this suppression is not sufficient. However, the method introduced so far suppresses FCNC to the level $O(G\alpha m_q^2/M_W^2)$, where m_q is the mass of the quark in the intermediate state. We can see this by considering the upper line of the box diagram shown in Fig. 9.1. The V_{KM} matrix elements which occur in this line include a mass-independent term,

$$V_{ij}^* V_{kJ} = V_{kj} V_{ji}^+ - \delta_{ki}, \tag{9.22}$$

and a mass-dependent term

$$V_{ij}^{*}V_{kj}\,\frac{m_{\mathrm{q}j}}{M_{\mathrm{W}}}=V_{kj}V_{ji}^{+}\,\frac{m_{\mathrm{q}j}}{M_{\mathrm{W}}}. \tag{9.23}$$

The leading term vanishes by virtue of the unitarity of the mixing matrix and the next term is proportional to the mass of the intermediate quark. This is the famous Glashow–Iliopoulos–Maiani cancellation scheme (Glashow *et al.*, 1970). Processes involving quarks in intermediate states, which are light relative to M_{W}, give a very small contribution. The mechanism has important consequences for box and penguin diagrams. These diagrams occur for K^0–$\bar{\mathrm{K}}^0$ mixing and the ε_{K} parameter.

The above requirements for flavor conservation in neutral couplings determines to a large extent the representation assignment of the fermion fields. There is a general theorem, which states that, for a gauge theory based on the group SU(2) × U(1), the bounds of FCNC are satisfied if we classify (Paschos, 1977; Glashow and Weinberg, 1977) the quarks into representations of the group in such a way that quarks of the same charge and the same helicity have the same T (total weak isospin) and T_3 (third component of isospin). For quarks of only two charges (2/3, −1/3) it implies that there must be equal numbers of up and down quarks.

We illustrate the implications of the result with some examples.

Example 1 Models with three quarks are not allowed, since they will produce strangeness-changing neutral currents. To solve this problem Glashow, Iliopoulos, and Maiani (Glashow *et al.*, 1970) introduced a charmed quark. The matrix M is

$$M=\begin{pmatrix}\cos\theta & \sin\theta\\ -\sin\theta & \cos\theta\end{pmatrix} \tag{9.24}$$

and the charged current

$$J_{\mu}^{+}=\left(\bar{\mathrm{u}}\,\bar{\mathrm{c}}\,\bar{\mathrm{d}}\,\bar{\mathrm{s}}\right)\begin{pmatrix}0 & M\\ 0 & 0\end{pmatrix}\begin{pmatrix}\mathrm{u}\\ \mathrm{c}\\ \mathrm{d}\\ \mathrm{s}\end{pmatrix}. \tag{9.25}$$

Example 2 Models with five quarks, u, c, d, s, and b, produce flavor-changing couplings. To eliminate these couplings the top quark was introduced. The charged-current interactions are now described by Eqs. (9.13) and (9.14).

9.3 The elements of the mixing matrix

There are many processes that determine values for the elements of the mixing matrix. They involve products of the weak couplings times hadronic matrix elements. Estimates of the latter require methods of strong interactions as they apply to low or high energies. For the sake of brevity, the description of hadronic methods given here is short. The aim is to give a general impression of the methods and arrive quickly at the relevant numerical results. The interested student will find more details in the references or in the following chapters, especially Chapters 11–14.

9.3.1 Determination of V_{ud}

For the weak interaction we introduced in Chapter 2 the Fermi coupling constant G_F, which is related to the SU(2) coupling by Eq. (8.14). It was also mentioned there that its numerical value is determined by the muon lifetime. To obtain a precise value for G_F it is necessary to include radiative corrections from the exchange of photons and gauge bosons, as well as the emission of photons. Such diagrams in general introduce infinities, which must be treated with special care. The electroweak theory is renormalizable and the infinities can be absorbed into a few coupling constants. In this book we do not cover the method of renormalization, but refer to an article and a book (Sirlin, 1978; Bardin and Passarino, 1999). The precise value for the Fermi coupling constant is

$$G_\mu = (1.166\,32 \pm 0.000\,04) \cdot 10^{-5}\,\mathrm{GeV}^{-2}, \tag{9.26}$$

with the subscript indicating that it is obtained from the muon lifetime.

The method which was used to construct the hadronic Lagrangian requires that the charged currents for quarks have the same coupling constant multiplied by the quark mixing matrix, as is seen in Eq. (9.14). This property is called universality. Thus the V_{ud} coupling is given by the ratio of the coupling constant measured in β-decay, to be denoted by G_V, to the muon decay constant:

$$V_{ud} = \frac{G_V}{G_\mu}. \tag{9.27}$$

The most accurate experiments for β-decay, so far, were done in nuclei and involve $0^+ \to 0^+$ transitions, also known as superallowed transitions. Their measurements and analyses have a long history. Precise determination of G_V must include radiative and in addition nuclear corrections. It is beyond the scope of this chapter to describe the corrections in detail. The result of the analyses is a very precise value,

$$V_{ud} = 0.9740 \pm 0.0003 \pm 0.0015, \tag{9.28}$$

where the first error is statistical and the second represents the theoretical uncertainty.

There are two other elementary transitions that are also relevant. The first is pion β-decay,

$$\pi^+ \to \pi^0 + e^+ + \nu.$$

The branching ratio for this decay has been measured to be

$$\mathrm{Br}\big(\pi^+ \to \pi^0 + e^+ + \nu\big) = (1.025 \pm 0.034) \times 10^{-8}, \tag{9.29}$$

which has a 3% error and is not as accurate as ratios from nuclear β-decays. The determination of V_{ud} from this decay has no nuclear corrections but carries a larger statistical error,

$$V_{ud} = 0.965 \pm 0.016. \tag{9.30}$$

The second elementary transition is the decay of neutrons: $n \to p + e^- + \bar{\nu}$. This decay depends both on the vector current and on the axial current, in contrast to the previous two cases, to which only the vector current contributes. Precise measurements of the neutron lifetime,

$$\tau_n = 888.5 \pm 0.8\,\mathrm{s}, \tag{9.31}$$

give the value

$$V_{ud} = 0.9801 \pm 0.0030. \tag{9.32}$$

We note that the three determinations are consistent with each other. The most accurate one from the superallowed nuclear transitions will be used later on.

9.3.2 Determination of V_{us}

There are several ways to determine V_{us}, among which the K_{l_3} decays are the cleanest. Hyperon decays give values that are almost as accurate. For the K-meson decays we use the reactions

$$K_L^0 \to \pi^- e^+ \nu_e \quad \text{and} \quad K^+ \to \pi^0 e^+ \nu_e.$$

The transition is from a pseudoscalar to another pseudoscalar particle and only the vector current contributes. Its matrix element can be written as

$$\langle \pi(p')|J_\mu|K(p)\rangle = C\big[\big(p_\mu + p'_\mu\big)f_+(q^2) + \big(p_\mu - p'_\mu\big)f_-(q^2)\big], \tag{9.33}$$

where $q^2 = (p - p')^2$, C is an isospin Clebsch–Gordan coefficient, and $f_\pm(q^2)$ are the form factors. In the SU(3)-symmetry limit at $q^2 = 0$ the form factor is known: $f_+(0) = 1$. Corrections to this value were computed to account for symmetry-breaking effects. Then the width

$$\Gamma = \frac{G_\mu^2 M_K^5}{192\pi^3} C^2 |f_+(0)|^2 |V_{us}|^2 \tag{9.34}$$

determines V_{us}. The result including all corrections is

$$V_{us} = 0.220 \pm 0.002. \tag{9.35}$$

The analysis of hyperon decays studies the decays of many hyperons, for which a χ^2 fit is performed. They involve form factors of both vector and axial currents, which are assumed to satisfy the SU(3) symmetry. The general fit is an impressive success of the SU(3) symmetry since there is a single value V_{us} consistent with all the hyperon data and the above value. It carries a slightly larger error arising from the theoretical uncertainties.

9.3.3 Determination of $|V_{cd}|$ and $|V_{cs}|$

One way to obtain the couplings $|V_{cd}|$ and $|V_{cs}|$ is to study the production of charmed particles in deep inelastic neutrino–nucleon scattering. The particles produced decay semileptonically and appear as events with opposite-sign dimuons. The elementary interactions are

$$\begin{aligned} &\nu + d \to \mu^- + c \\ &\qquad\qquad\quad \searrow \mu^+ + \nu + s, \end{aligned} \tag{9.36}$$

$$\begin{aligned} &\bar{\nu} + \bar{d} \to \mu^+ + \bar{c} \\ &\qquad\qquad\quad \searrow \mu^- + \bar{\nu} + \bar{s}. \end{aligned} \tag{9.37}$$

The semileptonic decays involve a mixture of charmed particles whose branching ratio is taken as $B_e = 7.1 \pm 1.3\%$.

Thus it remains to compute the production rates for charm quarks, which are discussed in Chapter 11. Here we mention that the original reactions are computed in the parton model as follows

$$\sigma(\nu N \to cX) = \frac{G^2 ME}{\pi}\left[r_d(U+D)|V_{cd}|^2 + 2r_s S|V_{cs}|^2\right], \tag{9.38}$$

$$\sigma(\bar{\nu} N \to \bar{c}X) = \frac{G^2 ME}{\pi}\left[r_{\bar{d}}(\bar{U}+\bar{D})|V_{cd}|^2 + 2r_{\bar{s}}\bar{S}|V_{cs}|^2\right]. \tag{9.39}$$

The r-coefficients measure the suppressions in the production of charmed quarks due to phase-space restrictions. Estimates for the experiments at energies $E_\nu = 220$ GeV and $E_{\bar{\nu}} = 150$ GeV gave the values

$$\begin{aligned} r_d\,(220\text{ GeV}) &= 0.91, & r_s\,(220\text{ GeV}) &= 0.72, \\ r_{\bar{d}}\,(150\text{ GeV}) &= 0.70, & r_{\bar{s}}\,(150\text{ GeV}) &= 0.66. \end{aligned} \tag{9.40}$$

The capital letters $U, D, \ldots$ denote integrals of the quark distribution functions in the proton. They are extracted from data on high-energy neutrino–nucleon scattering.

In Eqs. (9.38) and (9.39) we have two equations with two unknowns and thus we can solve for $V_{\rm cd}$ and $V_{\rm cs}$. The results are

$$|V_{\rm cd}| = 0.22 \pm 0.03, \tag{9.41}$$

$$|V_{\rm cs}| \geq 0.75. \tag{9.42}$$

Another source of information on $|V_{\rm cs}|$ is the semileptonic D-meson decays. They are proportional to $f_+^{\rm D\to K}(q^2)|V_{\rm cs}|$ and require estimates of the form factor $f_+^{\rm D\to K}(q^2)$.

The values for $|V_{\rm cs}|$ obtained by these methods have large errors due to theoretical uncertainties. These values have been superseded by measurements of W decays to identified charmed hadrons and the subsequent decays. W bosons decay to the pairs $({\rm u\bar{q}})$ and $({\rm c\bar{q}})$ with $\bar{\rm q} = \bar{\rm d},\ \bar{\rm s}, \bar{\rm b}$ antiquarks. The sum of the squares of the couplings for the six decays should add up to the value of 2. Since five of the six couplings are well measured or they are very small, LEP measurements can be converted into a precise value of

$$|V_{\rm cs}| = 0.996 \pm 0.016.$$

Without the use of unitarity the central value from all measurements is consistent, with 0.97 ± 0.10. With these values the upper-left-hand corner of the CKM matrix is known to a high degree of accuracy.

9.3.4 B-Meson decays and the determination of $V_{\rm cb}$ *and* $V_{\rm ub}$

The relatively long-lived B mesons made possible the determination of two more elements in the mixing matrix, $V_{\rm cb}$ and $V_{\rm ub}$. The decays of the B mesons proceed in the spectator model with the decay of b quarks into c and u quarks. The total width is the incoherent sum of the contributions from the above two decays, corrected, of course, for the exchange of gluons as described by QCD. The method accounts for the semileptonic and non-leptonic decays. Both decays were used in determining $V_{\rm cb}$, while the semileptonic spectrum is used for constraining the element $V_{\rm ub}$. In these estimates theoretical uncertainties enter the calculation and we shall discuss them in some detail.

The total width is given by

$$\Gamma_{\rm tot} = \Gamma_0(r|V_{\rm ub}|^2 + s|V_{\rm cb}|^2), \tag{9.43}$$

where r and s are products of phase space, color factors, and QCD corrections, and

$$\Gamma_0 = \frac{G^2 m_b^5}{192\pi^3}. \tag{9.44}$$

There are theoretical uncertainties for r, s, and Γ_0. For instance, the factor Γ_0 is sensitive to the b-quark mass:

$$\frac{1}{\Gamma^0} = \begin{cases} 0.93 \times 10^{-14}\,\mathrm{s} & \text{for} \quad m_b = 5.00\,\mathrm{GeV}, \\ 1.22 \times 10^{-14}\,\mathrm{s} & \text{for} \quad m_b = 4.75\,\mathrm{GeV}. \end{cases} \tag{9.45}$$

The spectator and parton models were used for analyzing, along these lines, the lepton spectra of semileptonic decays. A consistent analysis determines two more matrix elements, V_{cb} and V_{ub}, with $\approx$20% error.

An alternative method considers exclusive B decays in the heavy-quark effective theory (HQET). This is a systematic expansion in inverse powers of the heavy-quark mass. When the mass of the heavy quark is taken to infinity, the decays $B \to D^*\ell\bar{\nu}$ and $B \to D\ell\nu$ become equal. Eperimentally the two branching ratios are different, so corrections of $\mathcal{O}(1/m_q)$ must be included. Consequently, specific final states are selected to determine

$$|V_{cb}| = 0.041 \pm 0.002 \tag{9.46}$$

and

$$|V_{ub}| = 0.004 \pm 0.001. \tag{9.47}$$

Some details for the calculations are included in Section 14.5. Finally, the discovery of the top quark was achieved by observing semileptonic decays that provide an approximate estimate of $|V_{tb}|$.

With the B-meson decays we close the discussion concerning the elements of the CKM matrix, of which six elements are directly determined by experiments. Values for the remaining three elements, involving couplings of the top quark, are deduced from the unitarity of the matrix.

9.3.5 Summary and unitarity

In this chapter we were able to derive accurate values for the matrix elements from tree-level constraints. We emphasize that we can determine only their magnitudes, not their relative phases. In summary,

$$\begin{aligned} |V_{ud}| &= 0.9740 \pm 0.0020, & |V_{us}| &= 0.220 \pm 0.002, \\ |V_{cd}| &= 0.22 \pm 0.03, & |V_{cs}| &= 0.97 \pm 0.10, \\ |V_{cb}| &= 0.041 \pm 0.002, & |V_{ub}| &= 0.004 \pm 0.001. \end{aligned} \tag{9.48}$$

The unitarity of the mixing matrix restricts the matrix elements even further. The tree constraints together with unitarity give the following ranges reported by the Particle Data Group (Gilman *et al.*, 2002):

$$|V_{ij}| = \begin{pmatrix} 0.9741\text{–}0.9756 & 0.219\text{–}0.226 & 0.0025\text{–}0.0048 \\ 0.219\text{–}0.226 & 0.9732\text{–}0.9748 & 0.0038\text{–}0.004 \\ 0.004\text{–}0.014 & 0.037\text{–}0.044 & 0.9990\text{–}0.9993 \end{pmatrix}. \tag{9.49}$$

The small values for many of the elements justify the small-angle approximation and the Wolfenstein parametrization described in Section 9.1. Similarly, one obtains values for the other parametrizations as was done in Eq. (9.19). The phase δ is still undetermined. Its determination requires measurements of the CP parameters, which we postpone until Chapters 15 and 16.

We are now in a position to test the unitarity of the mixing matrix. There are two types of constraints.

(i) The sum of the squares of absolute values of the elements for each row or each column must sum up to unity. This can be tested for the first row,

$$|V_{\rm ud}|^2 + |V_{\rm us}|^2 + |V_{\rm ub}|^2 = 0.9970 \pm 0.0036, \tag{9.50}$$

which is consistent with unity. The radiative corrections for the $V_{\rm ud}$ element are very crucial because without them the right-hand side in (9.50) would be greater than unity, in fact $\sum_i |V_{{\rm u}i}|^2 = 1.020 \pm 0.004$.

(ii) A convenient and pictorial way to summarize the content of the CKM matrix is in terms of unitarity triangles. Consider the entries of each row or column of the matrix as the components of a vector. Then the unitarity condition applied to any two columns is the dot product of one column with the complex conjugate of another column. For the first and third columns the condition yields

$$V_{\rm ud} V_{\rm ub}^* + V_{\rm cd} V_{\rm cb}^* + V_{\rm td} V_{\rm tb}^* = 0. \tag{9.51}$$

The unitarity triangle is a geometrical representation of this equation in the complex plane. Each term in the equation is proportional to $A\lambda^3$ and, to leading order,

$$V_{\rm cd} V_{\rm cb}^* \approx -A\lambda^3, \qquad V_{\rm ud} V_{\rm ub}^* \approx A\lambda^3(\rho + {\rm i}\eta), \quad \text{and} \quad V_{\rm td} V_{\rm tb}^* \approx A\lambda^3(1 - \rho - {\rm i}\eta). \tag{9.52}$$

We can choose to orient the triangle so that $V_{\rm cd}\, V_{\rm cb}^*$ lies on the x-axis and scale out the common factor $A\lambda^3$ which is of order 1%. Now the coordinates for the vertices are shown in Fig. 9.2. The angles α, β, and γ of the triangle are also referred to as ϕ_2, ϕ_1, and ϕ_3, respectively. It is evident from the construction of the triangle that β and γ are the phases of the elements $V_{\rm td}$ and $V_{\rm ub}$, respectively:

$$V_{\rm td} = |V_{\rm td}|{\rm e}^{-{\rm i}\beta}, \quad V_{\rm ub} = |V_{\rm ub}|{\rm e}^{-{\rm i}\gamma}. \tag{9.53}$$

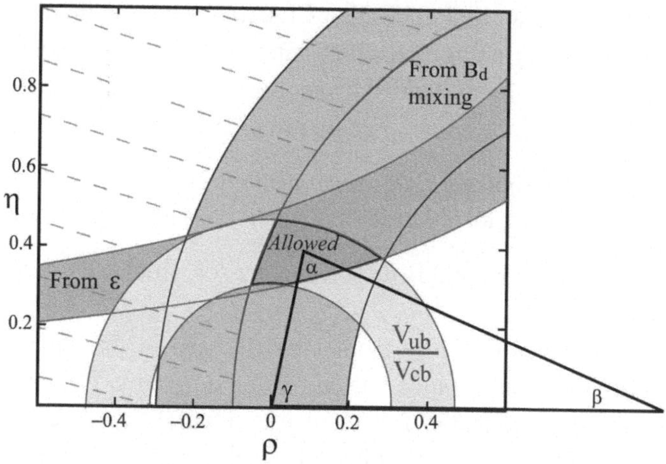

Figure 9.2. The unitarity triangle.

Once $A\lambda^3$ is factored out, the triangle depends on ρ and η. Let us select the x-axis to be ρ and the y-axis η. The shape of the triangle is now determined by three measurements. The first constraint comes from the magnitude of V_{ub} which determines a ring centered at the origin. The CP parameter ε_{K} determines a second region. Finally, the mass difference of the $\mathrm{B_d}$ mesons defines another ring; this time its center is at $\rho = 1$ (see Eqs. (15.53), (16.26) and (16.27)). The three regions are shown in Fig. 9.2, where their intersection defines the apex of the triangle. All additional measurements that depend on parameters of the triangle must reproduce the unitarity triangle (see Section 16.5).

We have mentioned already that CP violation is attributed to the phase in the CKM matrix. Quantitative predictions for CP asymmetries always contain

$$s_1 s_2 s_3 \sin\delta \qquad \text{or} \qquad s_\beta s_\gamma s_\theta \sin\delta' \tag{9.54}$$

as a multiplicative factor. For three generations of quarks there is a rephasing-invariant measure of CP violation. In terms of the elements, it is given by

$$J_{i\alpha} \equiv \mathrm{Im}\{V_{j\beta} V_{k\gamma} (V_{j\gamma} V_{k\beta})^*\}, \tag{9.55}$$

where i, j, k and α, β, γ are cyclic permutations of 1, 2, 3, i.e. once we give numerical values to i and α, the other indices are determined (Jarlskog, 1985). There are nine such invariants, which are all equal to each other. Their explicit form in the Maiani parametrization is

$$J_{i\alpha} \approx \beta\gamma s_\theta \sin\delta'. \tag{9.56}$$

For the central values of the angles

$$J_{i\alpha} \approx 2.5 \times 10^{-4} s_\theta \sin\delta'. \tag{9.57}$$

The smallness of this quantity implies that CP parameters will in general be small. There are also exceptions to this rule, which happen when the CP-violating quantity, which is proportional to $J_{i\alpha}$, is divided by another small quantity. It is evident from this discussion that the CP asymmetries manifest themselves in two ways:

(i) processes in which the rates are large have small asymmetries; and
(ii) large asymmetries occur for observables when the branching ratios are small.

Both situations appear in K- and B-meson decays, for which CP asymmetries have been observed.

Beyond the estimates of CKM elements discussed in this chapter, there are additional limits from observations related to loop diagrams. The theoretical analyses are now more complicated and involve additional theoretical assumptions. The advantage, however, is that they investigate the quantum nature of the theory and lead to a consistent picture. In fact, there are additional checks for the angles (CP phases) of the unitarity triangle. We shall cover several of these exciting topics in later chapters of the book.

References

Bardin, D., and Passarino, G. (1999), *The Standard Model in the Making* (Oxford, Oxford Science Publications)
Cabibbo, N. (1963), *Phys. Rev. Lett.* **10**, 531
Glashow, S. L., Iliopoulos, J., and Maiani, L. (1970), *Phys. Rev.* **D2**, 1285
Glashow, S. L., and Weinberg, S. (1977), *Phys. Rev.* **D15**, 1958
Jarlskog, C. (1985), *Phys. Rev. Lett.* **55**, 1039
Kobayashi, M., and Maskawa, T. (1973), *Prog. Theor. Phys.* **49**, 652
Paschos, E. A. (1977), *Phys. Rev.* **D15**, 1966
Sirlin, A. (1978), *Rev. Mod. Phys.* **50**, 573

Select bibliography

More details on determinations of matrix elements and additional references appear in

Gilman, F., Kleinknecht, K., and Renk, B. (2002), *Phys. Rev.* **D66**, 010001–13
Paschos, E. A., and Türke, U. (1989), *Phys. Rep.* **178**, 145
Kleinknecht, K. (2003), *Uncovering CP Violation* (Berlin, Springer-Verlag)

Part III

Experimental consequences and comparisons

10

Deep inelastic scattering

10.1 Kinematics for deep inelastic scattering

The processes that we are studying in the next few sections are shown schematically in Fig. 10.1. An initial neutrino with energy E hits a proton, producing a final state of a muon with energy E' and an undetected final hadronic state.

The lepton vertex is well known. All the interesting structure is included in the hadronic vertex. The kinematics are shown in the diagram and they involve

k_μ, the four-vector of the neutrino,
k'_μ, the four-vector of the muon,
$q = k - k'$, the four-momentum transferred from leptons to hadrons,
P_μ, the four-momentum of the target nucleon,
$\nu = pq/M$, energy transfer in the laboratory frame,
θ, the laboratory angle of the muon produced relative to the incident neutrino, and
$Q^2 = -q^2 = -m_\mu^2 + 2E\vec{k}\,'(1 - \cos\theta) \approx 4EE'\sin^2\theta/2$, with $k' = \sqrt{E'^2 - m_\mu^2}$.

The above definitions hold also for electroproduction, when the initial neutrino is replaced by an electron and the exchange particle is the photon. The discussion of this and the following section is restricted to neutrino reactions. The cross section for such a process in the rest frame of the proton is given by

$$d\sigma = \frac{1}{(2E)(2M)} \sum\!\!\!\!\!\!\int_n |\mathcal{M}|^2 (2\pi)^4 \delta^4(k + p - k' - p_n) \frac{d^3k'}{2E'(2\pi)^3}, \qquad (10.1)$$

where

$$p_n^\mu = \sum_{i=1}^{n} p_i^\mu,$$

with the summation over all final particle configurations, each of which contains n particles with momenta p_i and $i - 1, \ldots, n$. The integration over the phase space

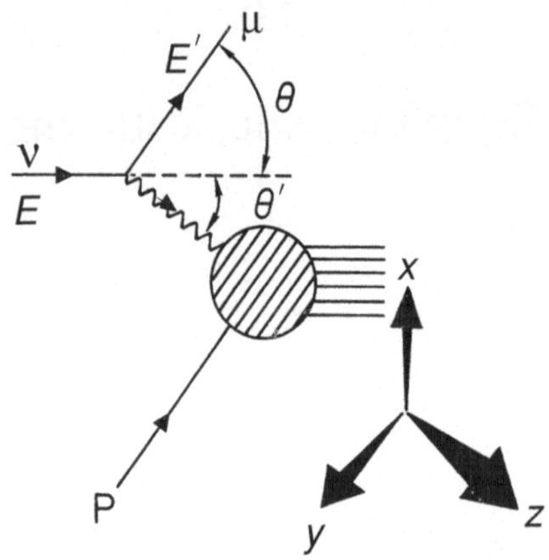

Figure 10.1. Inelastic neutrino–nucleon scattering, together with the coordinate system used in decomposing the leptonic current.

of final-state particles and the summation over the configurations is given by

$$\sum\!\!\!\!\!\!\!\int_n \cdots = \sum_n \int \prod_{i=1}^{n} \left[\frac{\mathrm{d}^3 p_i}{2E_i(2\pi)^3} \right] \cdots$$

Later on we shall specify the final state to be a single quark, with the product reduced to a single phase-space factor.

The matrix element is

$$\mathcal{M} = \frac{G}{\sqrt{2}} \bar{u}(k')\gamma_\mu(1-\gamma_5)u(k)\langle p_n|J^\mu|p\rangle. \tag{10.2}$$

We write the leptonic current as

$$j_\mu^{\text{lept}} = \bar{u}(k')\gamma_\mu(1-\gamma_5)u(k). \tag{10.3}$$

Neglecting the muon mass, the current is evaluated by multiplying it by a simple factor (Bjorken and Paschos, 1970):

$$j_\mu^{\text{lept}} = \sum_{s,s'} \bar{u}(k',s')\gamma_\mu(1-\gamma_5)u(k,s) \frac{\bar{u}(k,s)\gamma_0(1-\gamma_5)u(k',s')}{\bar{u}(k,s)\gamma_0(1-\gamma_5)u(k',s')}. \tag{10.4}$$

The factor of unity is introduced in order to change the numerator into a trace; the summation over spins does not change the lepton current because $(1-\gamma_5)$ is a chirality-projection operator, so the extra states introduced by $\sum_{s,s'}$ contribute zero:

$$\begin{aligned} j_\mu^{\text{lept}} &= \frac{2\,\mathrm{Tr}\big[\gamma_\mu(1-\gamma_5)\not{k}\,\gamma_0\,\not{k}'\big]}{\{2\,\mathrm{Tr}[\gamma_0(1-\gamma_5)\not{k}'\gamma_0\,\not{k}]\}^{1/2}} \\ &= \frac{8\big(k_\mu E' + k'_\mu E - g_{\mu o}\,k\cdot k' + \mathrm{i}\varepsilon_{\mu o\alpha\beta}k^\alpha k'^\beta\big)}{4\sqrt{EE'}\cos\theta/2}. \end{aligned} \tag{10.5}$$

Similarly, we calculate the square of the denominator, which produces a trace.

Using current conservation, we can eliminate one of the components in j_μ and expand the current in terms of three orthonormal polarization vectors whose spatial components lie along the axes shown in Fig. 10.1; the z-axis lies along q. This decomposition simplifies considerably in the high-energy limit $\nu \gg 2M \approx 2\,\text{GeV}$; $Q^2 \ll \nu^2$, which is all we consider in this chapter. An alternative way would be to square the leptonic current and compute the leptonic tensor. The method is straightforward and the interested student can use it in order to reproduce some of the formulas in Sections 10.2 and 10.3. Here we find the method convenient for introducing helicity cross sections.

The three polarization vectors below correspond to the angular-momentum state $|J = 1, m\rangle$ with helicities $m = 0$, 1, and -1, respectively:

$$\begin{aligned} \varepsilon_\mu^{\text{S}} &= \frac{1}{\left[Q^2\right]^{1/2}}(q_z,\ 0,\ 0,\ q_0) \approx \frac{\nu}{\left[Q^2\right]^{1/2}}\left(1 + \frac{Q^2}{2\nu^2},\ 0,\ 0,\ 1\right), \\ \varepsilon_\mu^{\text{R}} &= \frac{1}{\sqrt{2}}(0,\ 1,\ \text{i},\ 0), \\ \varepsilon_\mu^{\text{L}} &= \frac{1}{\sqrt{2}}(0,\ 1,\ -\text{i},\ 0). \end{aligned} \tag{10.6}$$

They satisfy the conditions $\varepsilon_{\text{S}}^2 = +1$, $|\varepsilon_{\text{L;R}}|^2 = -1$, and $\varepsilon_{\text{S,L,R}} \cdot q = 0$. In the high-energy approximation the current, evaluated in the laboratory frame, becomes

$$j_\mu^\ell \approx 4\frac{(EE'Q^2)^{1/2}}{\nu}\left[\varepsilon_\mu^{\text{S}} + \left(\frac{E'}{2E}\right)^{\frac{1}{2}}\varepsilon_\mu^{\text{R}} + \left(\frac{E}{2E'}\right)^{\frac{1}{2}}\varepsilon_\mu^{\text{L}}\right]. \tag{10.7}$$

The only change in j_μ^{lept} on going over to antineutrino-induced processes is the interchange $\text{R} \leftrightarrow \text{L}$.

The integration over the phase space of the muon can be carried out,

$$\frac{\text{d}^3k'}{2E'(2\pi)^3} = \frac{E'\,\text{d}E'\,\text{d}\Omega}{2(2\pi)^3}.$$

In addition, we can transform to an invariant phase-space element,

$$\frac{\text{d}\sigma}{\text{d}Q^2\,\text{d}\nu} = \frac{\pi}{EE'}\frac{\text{d}\sigma}{\text{d}\Omega\,\text{d}E'},$$

arriving at

$$\frac{\text{d}\sigma}{\text{d}Q^2\,\text{d}\nu} = \frac{G^2}{2\pi^2}\frac{E'}{E}\frac{Q^2}{\nu}\left(\frac{1}{(2\nu)(2M)}\sum\!\!\!\!\!\!\int_n |\langle n|\tilde{j}_\mu J^\mu|p\rangle|^2(2\pi)^4\delta^4(p_n - p - q)\right). \tag{10.8}$$

Here

$$\tilde{j}_\mu^{\,\text{lept}} = \varepsilon_\mu^{\text{S}} + \left(\frac{E'}{2E}\right)^{\frac{1}{2}} \varepsilon_\mu^{\text{R}} + \left(\frac{E}{2E'}\right)^{\frac{1}{2}} \varepsilon_\mu^{\text{L}}.$$

It is evident now that the amplitude $\langle n|\tilde{j}_\mu J^\mu|p\rangle$ is the sum of three helicity amplitudes: scalar (A_{S}), right-handed (A_{R}), and left-handed (A_{L}). The cross section is the sum of three helicity cross sections and three interference terms. When we average over the azimuthal angles of the hadrons produced, the three interference terms average to zero, as indicated by the following argument.

Let Γ be a fixed set of final-state hadron momenta that are measured. Let $\Gamma' = R\Gamma$ be the set of momenta obtained by rigid rotation of Γ about $\vec{q}$ (the z-axis) by the angle ϕ. We kept the neutrino and muon momenta fixed and rotated the hadronic system. This is equivalent to keeping the hadrons fixed and rotating the neutrino–muon plane in the opposite direction. Under this rotation the only change in the cross section is to replace $\tilde{j}_\mu^{\,\text{lept}}$ as follows:

$$\tilde{j}_\mu^{\,\text{lept}} = \varepsilon_\mu^{\text{S}} + \sqrt{\frac{E'}{2E}}\, \varepsilon_\mu^{\text{R}} \mathrm{e}^{\mathrm{i}\phi} + \sqrt{\frac{E}{2E'}}\, \varepsilon_\mu^{\text{L}} \mathrm{e}^{-\mathrm{i}\phi}. \tag{10.9}$$

The rotation is equivalent to a rotation of the two polarization vectors $\vec{\varepsilon}_{\text{R,L}}$ around the z-axis. $\vec{\varepsilon}^{\,\text{S}}$, which is parallel to $\vec{q}$, does not change. Accordingly, only the interference terms $2\Re(A_{\text{S}}A_{\text{R}}^*\mathrm{e}^{\mathrm{i}\phi})$, $2\Re(A_{\text{S}}A_{\text{L}}^*\mathrm{e}^{-\mathrm{i}\phi})$, and $2\Re(A_{\text{R}}A_{\text{L}}^*\mathrm{e}^{2\mathrm{i}\phi})$ change. They produce terms linear in $\cos(\phi)$ and $\sin(\phi)$. By averaging over the azimuthal orientations of the final hardons, i.e. integrating over ϕ from 0 to 2π, the interference terms are made to vanish. Should one wish to isolate the interference terms, then it is necessary to construct appropriate moments over the angle ϕ.

To sum up, Γ denotes a set of hadronic momenta in the final state, whose angles relative to each other are kept fixed; the rigid rotation around $\vec{q}$ has been averaged, i.e. integrated out. In this manner only helicity cross sections survive.

We define the helicity cross sections for absorption of the "virtual" W nucleon into final hadronic states by

$$\sigma^{(\lambda)}(\nu, Q^2) = \frac{1}{(2\nu)(2M)} \int |\langle n|\varepsilon_\mu^\lambda \cdot J^\mu(0)|p\rangle|^2 (2\pi)^4 \delta^{(4)}(p' - p - q) \frac{\mathrm{d}^3 p'}{2E_{p'}(2\pi)^3}. \tag{10.10}$$

Here we have assumed that there is only one particle in the final hadronic state. When there are many particles produced, the phase space is replaced by a product of phase-space factors. These cross sections depend only on ν and Q^2. The final formula reads

$$\frac{\mathrm{d}\sigma}{\mathrm{d}Q^2\,\mathrm{d}\nu} = \frac{G^2}{4\pi^2} \frac{E'}{E} \frac{Q^2}{\nu} \left(2\sigma_{\text{S}} + \frac{E'}{E}\sigma_{\text{R}} + \frac{E}{E'}\sigma_{\text{L}}\right). \tag{10.11}$$

The absorption cross sections are not uniquely defined; for $q^2 = 0$ the flux factor for the exchanged particle is 2ν. This is a convention and sometimes the factor has been replaced by $2\nu[1 - Q^2/(2M\nu)]$. To avoid the zeros which appear for elastic scattering, we chose the overall factor in the helicity cross sections $F = 4M\nu$ (Eq. (10.10))

On introducing the structure function

$$W_2(\nu, Q^2) = \frac{1}{2\pi}\frac{Q^2}{\nu}(2\sigma_S + \sigma_R + \sigma_L) \tag{10.12}$$

and the ratios

$$(L) = \frac{\sigma_L}{2\sigma_S + \sigma_R + \sigma_L} \leq 1,$$
$$(R) = \frac{\sigma_R}{2\sigma_S + \sigma_R + \sigma_L} \leq 1, \tag{10.13}$$

it follows that

$$\frac{d\sigma}{dQ^2\, d\nu} = \frac{G^2}{2\pi}\frac{E'}{E} W_2(\nu, Q^2)\left[1 + \frac{\nu}{E'}(L) - \frac{\nu}{E}(R)\right]. \tag{10.14}$$

The antineutrino–nucleon cross section is obtained from (10.14) by the interchange L $\leftrightarrow$ R. We shall find these cross sections useful in several applications later on.

An alternative notation introduces the structure functions $W_1(\nu, Q^2)$, $W_2(\nu, Q^2)$, and $W_3(\nu, Q^2)$ defined in the next section. They are related to the absorption cross sections through (10.12) and the following:

$$W_1(\nu, Q^2) = W_2(\nu, Q^2)\left(1 + \frac{\nu^2}{Q^2}\right)[(L) + (R)], \tag{10.15}$$

$$W_3(\nu, Q^2) = W_2(\nu, Q^2)\frac{2M}{Q}\left(1 + \frac{\nu^2}{Q^2}\right)^{\frac{1}{2}} [(L) - (R)]. \tag{10.16}$$

In the limit $\nu^2/Q^2 \gg 1$ they reduce to

$$W_1(\nu, Q^2) = \nu W_2(\nu, Q^2)\frac{\nu}{Q^2} [(L) + (R)], \tag{10.17}$$

$$\nu W_3(\nu, Q^2) = \nu W_2(\nu, Q^2)\frac{2M\nu}{Q^2} [(L) - (R)]. \tag{10.18}$$

10.2 Hadronic structure functions

In the previous section we introduced structure functions that describe the hadronic vertex. Here we describe their connection with products of currents and their commutators. The formalism of this section is convenient in discussing sum rules or the light-cone behavior of the product of weak currents. We define the hadronic tensor as

$$W_{\mu\nu} = (2\pi)^3 \overline{\sum_{S_n}} \sum\!\!\!\!\!\!\int_n \langle P|J_\mu^+(0)|P_n\rangle\langle P_n|J_\nu(0)|P\rangle\delta^{(4)}(p_n - p - q). \quad (10.19)$$

Here $\sum\!\!\!\!\!\!\int_n$ sums over final states, and $\overline{\sum}_{S_n}$ averages over the spins of the target nucleon. By exponentiating the delta function and using translation invariance,

$$J_\mu(x) = \mathrm{e}^{\mathrm{i}Px} J_\mu(0)\mathrm{e}^{-\mathrm{i}Px}, \quad (10.20)$$

one obtains

$$W_{\mu\nu} = \frac{1}{2\pi}\overline{\sum_{S_n}} \int \mathrm{d}^4x\, \mathrm{e}^{\mathrm{i}qx}\langle P|J_\mu^+(x)J_\nu(0)|P\rangle, \quad (10.21)$$

where the unitary relation $\sum\!\!\!\!\!\!\int_n |p_n\rangle\langle p_n| \equiv 1$ was used. We may change $W_{\mu\nu}$ into a commutator,

$$W_{\mu\nu} = \frac{1}{2\pi}\overline{\sum_{S_n}} \int \mathrm{d}^4x\, \mathrm{e}^{\mathrm{i}qx}\langle P|\big[J_\mu^+(x), J_\nu(0)\big]|P\rangle, \quad (10.22)$$

since the second term of the commutator,

$$\frac{1}{2\pi}\int \mathrm{d}^4x\, \mathrm{e}^{\mathrm{i}qx}\langle P|J_\nu(0)J_\mu^+(x)|P\rangle, \quad (10.23)$$

vanishes. This is proven by reversing the steps and showing that (10.22) reduces to

$$\sum\!\!\!\!\!\!\int_n \langle P|J_\nu(0)|p_n\rangle\langle p_n|J_\mu^+(0)|P\rangle\delta^{(4)}(p_n - P + q) = 0, \quad (10.24)$$

since in the physical process $q_0 = E_n - M_{\text{proton}} \geq 0$, but in Eq. (10.24) the δ-function argument implies

$$q_0 = P_0 - p_n^0 = M - E_n < 0.$$

By virtue of Lorentz and gauge invariance, $W_{\mu\nu}$ can be written in terms of six scalar functions, which are better known as structure functions. In neutrino scattering, however, only three contribute to the inelastic cross section because the lepton current is conserved (for $m_\mu = 0$). The tensor relevant to deep inelastic scattering is

$$W_{\mu\nu} = -g_{\mu\nu}W_1 + \frac{P_\mu P_\nu}{M^2}W_2 - \mathrm{i}\frac{\varepsilon_{\mu\nu\alpha\beta}P^\alpha q^\beta}{2M^2}W_3, \quad (10.25)$$

where the structure functions $W_1(Q^2, \nu)$ and $W_2(Q^2, \nu)$ arise from the product of vector $\otimes$ vector currents and axial $\otimes$ axial currents, whereas $W_3(Q^2, \nu)$ is the interference of an axial current ⊗ a vector current. The additional three terms are

$$\frac{q_\mu q_\nu}{M^2}W_4 + \frac{P_\mu q_\nu + P_\nu q_\mu}{M^2}W_5 + \mathrm{i}\frac{P_\mu q_\nu - P_\nu q_\mu}{M^2}W_6.$$

Their contributions to the matrix elements and the cross section are proportional to lepton masses and will be neglected.

10.3 Scaling and the total cross section

The structure functions are functions of ν and Q^2, but at high energies both variables are very large. It was suggested by Bjorken (1969) that, in the limit $\nu \to \infty$, $Q^2 \to \infty$, with the ratio

$$x = \frac{Q^2}{2M\nu} = \text{finite}, \tag{10.26}$$

the structure functions become functions of x only, i.e.

$$\nu W_{2,3}(\nu, Q^2) \to F_{2,3}(x), \tag{10.27}$$

$$M W_1(\nu, Q^2) \to F_1(x). \tag{10.28}$$

This was established in experiments on deep inelastic electron–proton scattering, for which the limit is reached at relatively low values of $Q^2, 2M\nu \approx (1\ \text{GeV})^2$. Inelastic electron–proton scattering is closely related to neutrino reactions and we mention it later on in this section.

In the scaling limit the relations (10.18) and (10.19) reduce to

$$\begin{aligned} 2xF_1(x) &= F_2(x)[(L) + (R)], \\ xF_3(x) &= F_2(x)[(L) - (R)]. \end{aligned}$$

With this notation we can rewrite the cross section in a convenient form. For variables we use $x = Q^2/(2M\nu)$ and the inelasticity $y = \nu/E$, then we substitute the scaling functions into the cross section, Eq. (10.14), and change the phase-space variables to arrive at

$$\frac{\mathrm{d}\sigma}{\mathrm{d}x\,\mathrm{d}y} = \frac{G^2ME}{\pi}\left[xy^2F_1(x) + (1-y)F_2(x) + xy\left(1 - \frac{1}{2}y\right)F_3(x)\right],$$

where the structure functions depend on the process under consideration. In order to obtain the corresponding cross section for an antineutrino-induced reaction, one should change the sign of the F_3 term and replace the structure functions with the charge conjugate.

The main difference between electroproduction and neutrino-induced reactions is the nature of the particle exchanged. In electroproduction the particle exchanged is the photon, which has only a vector coupling

$$j_\mu^{\text{lept}} = \bar{u}(k')\gamma_\mu u(k). \tag{10.29}$$

The vector–axial interference term is now absent and the cross sections σ_R and σ_L are equal. Following steps similar to those of the previous section, one finds

$$\frac{d\sigma(\text{ep})}{dQ^2\, d\nu} = \frac{E'}{E}\frac{4\pi\alpha^2}{Q^4}\left[W_2^{\text{e}}\cos^2\left(\frac{\theta}{2}\right) + 2W_1^{\text{e}}\sin^2\left(\frac{\theta}{2}\right)\right], \tag{10.30}$$

where W_1^{e} and W_2^{e} are electroproduction structure functions analogous to those introduced in Eq. (10.25). In the cross section we kept the scattering angle θ. However, we can substitute it in terms of Q^2 and the energies E and E' and arrive at a formula analogous to (10.14). The superscript e indicates their electromagnetic origin. Numerous experiments have shown that the limits

$$\nu W_2^{\text{e}}(\nu, Q^2) \to F_2^{\text{e}}(x), \tag{10.31}$$

$$M W_1^{\text{e}}(\nu, Q^2) \to F_1^{\text{e}}(x) \tag{10.32}$$

are reached for relatively low values of ν and Q^2. This is shown in Fig. 10.2, where the structure function $F_2(x)$ is plotted for a range of Q^2. Deviations from the scaling law have also been established, and we return to this topic in Chapter 11. We show next that scaling predicts $\sigma_{\text{tot}} \sim E_\nu$: namely a linear rise with neutrino energy.

From (10.14) and scale invariance (10.27) we find the averaged cross section over protons and neutrons

$$\begin{aligned}\frac{d\sigma}{d\nu} &= \frac{G^2}{2\pi}\frac{E'}{E}\int_{\sim 0}^{2M\nu}\frac{dQ^2}{\nu}\nu W_2(\nu, Q^2)\left(1 + \frac{\nu}{E'}(L) - \frac{\nu}{E}(R)\right)\\ &= \frac{G^2 M}{\pi}\frac{E'}{E}\left(1 + \frac{\nu}{E'}\langle L\rangle - \frac{\nu}{E}\langle R\rangle\right)\int_0^1 dx\,\frac{1}{2}\left[F_2(x)_\text{p} + F_2(x)_\text{n}\right],\end{aligned} \tag{10.33}$$

where $\langle R\rangle$ and $\langle L\rangle$ imply that the appropriate averages of x have been taken. Then the total cross section is

$$\sigma_{\text{tot}} = \frac{G^2 M E}{\pi}\int_0^1 dx\,\frac{1}{2}\left[F_2(x)_\text{p} + F_2(x)_\text{n}\right]\left\{\frac{1}{2} + \frac{\langle L\rangle}{2} - \frac{\langle R\rangle}{6}\right\}. \tag{10.34}$$

The factor in the curly brackets lies between 1 and $\frac{1}{3}$. In particular,

$$\frac{1}{2} + \frac{1}{2}\langle L\rangle - \frac{1}{6}\langle R\rangle = \begin{cases} 1 & \text{if} \quad \sigma_R = \sigma_S = 0,\\ \frac{2}{3} & \text{if} \quad \sigma_R = \sigma_L,\ \sigma_S = 0,\\ \frac{1}{2} & \text{if} \quad \sigma_R = \sigma_L = 0,\\ \frac{1}{3} & \text{if} \quad \sigma_L = \sigma_S = 0.\end{cases} \tag{10.35}$$

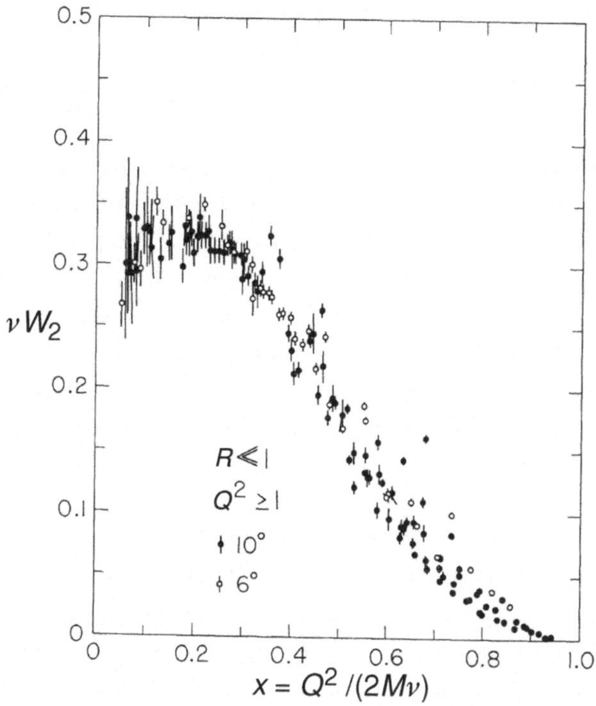

Figure 10.2. Scaling of the structure function $\nu W_2^e = F_2(x)$.

From (10.33) we see that a linear rise in σ_{tot} depends on the property that νW_2 is scale-invariant and the absence of a W propagator. The neutrino measurements give

$$\sigma_{tot}^{\nu} = (0.677 \pm 0.014) \times 10^{-38}\,\text{cm}^2\,\frac{E_\nu}{\text{GeV}}.$$

We can also compare neutrino and antineutrino cross sections on isoscalar targets:

$$\frac{\sigma^{\bar{\nu}N}}{\sigma^{\nu N}} = \frac{\frac{1}{2} + \frac{1}{2}\langle R\rangle - \frac{1}{6}\langle L\rangle}{\frac{1}{2} + \frac{1}{2}\langle L\rangle - \frac{1}{6}\langle R\rangle}.$$

The ratio is bounded between $\frac{1}{3}$ and 3. The experimental data give

$$\sigma_{tot}^{\bar{\nu}} = (0.334 \pm 0.008) \times 10^{-38}\,\text{cm}^2\,\frac{E_{\bar{\nu}}}{\text{GeV}},$$

with the ratio of the two slopes being 0.501 ± 0.015, which is consistent with the above prediction and close to the lower bound.

10.4 The parton model

"Friends rush in where angels fear to tread."

(R. P. Feynman, at Fermilab, 1973)

A physical interpretation of the scaling phenomenon is given by the parton model, which considers the scattering as the incoherent sum of scattering from point-like constituents within the proton, called partons. The point-like nature of the constituents reproduces scaling. By studying several reactions it was possible to deduce properties of the constituents, such as electric charge, and identify the partons with quarks. The parton model has been applied to a wide range of high-energy reactions, many of which will be covered in this chapter. Deep inelastic reactions together with hadron spectroscopy supply the major evidence for the quark substructure of matter.

Neutrino–nucleon scattering

The basic idea in the parton model is to regard the deep inelastic scattering as quasi-free scattering from point-like constituents within the proton. This happens when the scattering is viewed from a frame in which the proton has infinite momentum. The neutrino–proton center-of-mass system is, at high energies, a good approximation of such a frame. In the infinite-momentum frame, the proton is Lorentz-contracted into a thin pancake, and the lepton scatters instantaneously. Furthermore, the proper motion of the constituents within the proton is slowed down by time dilatation. We estimate the interaction time and lifetime of the virtual states within the proton. In the notation of the previous section and Fig. 10.3, the initial electron and proton are collinear and in opposite directions:

$$\begin{aligned} \vec{k} &= -\vec{P}, \\ k_0 &\approx P_0 = P. \end{aligned} \tag{10.36}$$

In this frame

$$p \cdot q = M\nu = (q_0 + q_z)P, \tag{10.37}$$

$$k \cdot q = -Q^2/2 = (q_0 - q_z)P, \tag{10.38}$$

from which it follows that

$$q_0 = \frac{2M\nu - Q^2}{4P}. \tag{10.39}$$

The time of interaction is $\tau \approx 1/q_0$, which for moderate values of x decreases as

$$\tau = \frac{4P}{2M\nu(1-x)}. \tag{10.40}$$

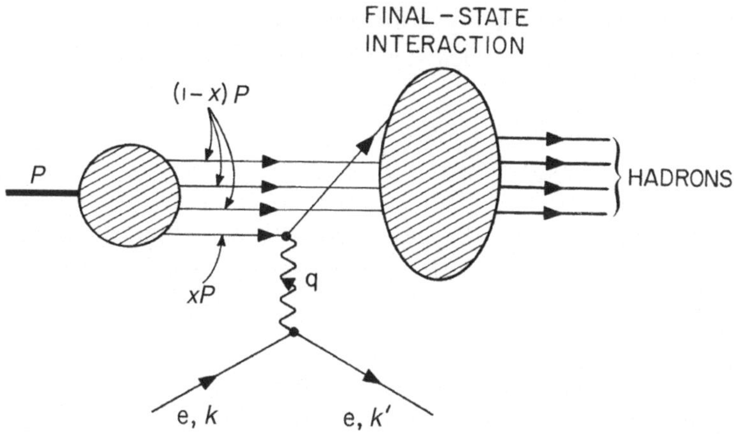

Figure 10.3. Kinematics for neutrino–nucleon scattering in the parton model.

We visualize the proton as composed of virtual states called partons. We denote by x the fraction of the proton's momentum carried by a constituent. The lifetime of the virtual states (Feynman, 1969; Bjorken and Paschos, 1969) is

$$T = \frac{1}{E_x + E_{1-x} - E_{\mathrm{p}}} = \frac{1}{\sqrt{(xP)^2 + \mu_1^2} + \sqrt{(1-x)^2P^2 + \mu_2^2} - \sqrt{P^2 + M^2}}$$
$$\approx \frac{2P}{\left(\mu_1^2 + P_{1\perp}^2\right)/x + \left(\mu_2^2 + P_{2\perp}^2\right)/(1-x) - M^2}. \tag{10.41}$$

If we now require that $\tau \ll T$, then we must consider the partons, contained in the proton, as free during the interaction. In this limit the current interacts with just one of the constituents, leaving the rest undisturbed, thus making the impulse approximation valid. The above conditions appear to be satisfied in high-energy and large-momentum-transfer electron–nucleon scattering and in high-energy neutrino–nucleon scattering. The model could fail for $x \to 0$ or 1, for which the expansion in (10.39) is no longer justified. The reader may have noticed that we use x with two meanings: the first one is Bjorken's variable x defined in Eq. (10.26) and the second is the fraction of the proton's momentum. This was done on purpose because the two variables are the same.

The cross section of a proton is the incoherent sum of cross sections of the individual constituents. We denote by $\mathrm{d}\sigma_i(x)/(\mathrm{d}Q^2\,\mathrm{d}\nu)$ the cross section of a neutrino on a parton of type i, which carries a fraction x of the proton's momentum,

$$p_i^\mu \approx xP^\mu. \tag{10.42}$$

We denote by $f_i(x)$ the probability of finding the ith constituent carrying a fraction x of the proton's momentum. Then the cross section is

$$\frac{d\sigma}{dQ^2\,d\nu} = \sum_i \int_0^1 \frac{d\sigma_i(x)}{dQ^2\,d\nu} f_i(x)dx. \tag{10.43}$$

The summation here is over all types of constituents within the proton and the integral is over the momentum fraction x. Thus the complicated hadronic structure is reduced to the incoherent scattering from point-like constituents times the structure functions $f_i(x)$.

The point cross sections have already been derived in Section 8.3. For neutrino–parton scattering

$$\frac{d\sigma}{dQ^2\,d(p_i q/m_i)} = \frac{G^2}{\pi}\delta\left(\frac{p_i q}{m_i} - \frac{Q^2}{2m_i}\right) \tag{10.44}$$

or

$$\frac{d\sigma}{dQ^2\,d(p_i q)} = \frac{G^2}{\pi}\delta\left(p_i q - \frac{Q^2}{2}\right). \tag{10.45}$$

For neutrino–antiparton scattering

$$\frac{d\sigma}{dQ^2\,d(p_i q)} = \frac{G^2}{\pi}\left(1 - \frac{p_i q}{p_i k^\mu}\right)^2 \delta\left(p_i q - \frac{Q^2}{2}\right), \tag{10.46}$$

with k^μ the four-momentum of the neutrino.

The proton is built from two up quarks and one down quark, which constitute the valence quarks and give the proton its quantum numbers. In addition, there is in the proton a cloud of quark–antiquark pairs produced by the radiation of gluons and their subsequent conversion into pairs. The number of the pairs is infinite, but their momentum distributions have not been calculated explicitly. We denote the probability of finding an up quark carrying a fraction x of the proton's momentum by $u(x)$. Similarly, we denote by $d(x)$ the probability of finding a down quark carrying a fraction x of the proton's momentum. The cloud of quark–antiquark pairs of any flavor necessitates the introduction of additional quark distribution functions. For instance, $\bar{u}(x)$ and $\bar{d}(x)$ correspond to up and down antiquarks. Similarly, there are distributions $s(x)$, $\bar{s}(x)$, $c(x)$, $\bar{c}(x)$, ... for strange, charm, and other flavors.

When we substitute $p_i^\mu = xP^\mu$ into (10.44) we obtain the point cross section

$$\frac{d\sigma_i}{dQ^2\,d\nu} = \frac{G^2}{\pi} Mx\delta\left(xM\nu - \frac{Q^2}{2}\right). \tag{10.47}$$

Finally, on substituting the point cross sections in (10.42) and integrating over x, we arrive at the neutrino–proton scattering

$$\frac{d\sigma^{\nu p}}{dQ^2\,d\nu} = \frac{G^2}{\pi}\frac{x}{\nu}\left[d(x) + \bar{u}(x)\left(1 - \frac{\nu}{E}\right)^2\right]. \tag{10.48}$$

Here we omit the contribution from the strange and heavier quarks and their antiparticles. In the case of antineutrino–proton scattering we obtain

$$\frac{d\sigma^{\bar{\nu}p}}{dQ^2\,d\nu} = \frac{G^2}{\pi}\frac{x}{\nu}\left[\bar{d}(x) + u(x)\left(1 - \frac{\nu}{E}\right)^2\right]. \tag{10.49}$$

It is now evident that the momentum fraction $x = Q^2/(2M\nu)$ is indeed the Bjorken scaling variable.

The general case with many families of quarks can be easily written down. The contribution from a quark $q(x)$ and an antiquark $\bar{q}(x)$ is

$$\frac{d\sigma^{\nu N}}{dx\,dy} = \frac{G^2}{\pi}2MEx\left[q(x) + (1 - y)^2\bar{q}(x)\right], \tag{10.50}$$

provided that the quark under consideration is allowed by charge conservation. Similarly, the antineutrino–nucleon cross section is

$$\frac{d\sigma^{\bar{\nu}N}}{dx\,dy} = \frac{G^2}{\pi}2MEx\left[(1 - y)^2 q(x) + \bar{q}(x)\right]. \tag{10.51}$$

These relations are used to determine the antiquark content of the proton. For instance, the antineutrino–nucleon cross section at $y = 1$ measures $\bar{q}$. The total cross sections are also easily derived. They grow linearly with neutrino or antineutrino energy.

Finally the ratio of the total cross sections for an isoscalar target, such as deuterium or oxygen, is

$$\frac{\sigma^{\bar{\nu}d}}{\sigma^{\nu d}} = \frac{\int_0^1 dx\,x\left[\frac{1}{3}(u + d) + (\bar{u} + \bar{d})\right]}{\int_0^1 dx\,x\left[(u + d) + \frac{1}{3}(\bar{u} + \bar{d})\right]}. \tag{10.52}$$

Taking the experimental ratio of the cross sections to be approximately 0.50, we arrive at the conclusion that the integrated antiquark contribution is approximately 20% of the quark contribution.

We close this section with a few remarks. We derived the general formulas for neutrino- and antineutrino-induced reactions using helicity cross sections. We have also shown explicitly that they are related to the structure functions. The formalism can be carried over to electroproduction, for which similar formulas hold. We also emphasized that high-energy neutrino reactions are closely related to electroproduction in the deep inelastic region.

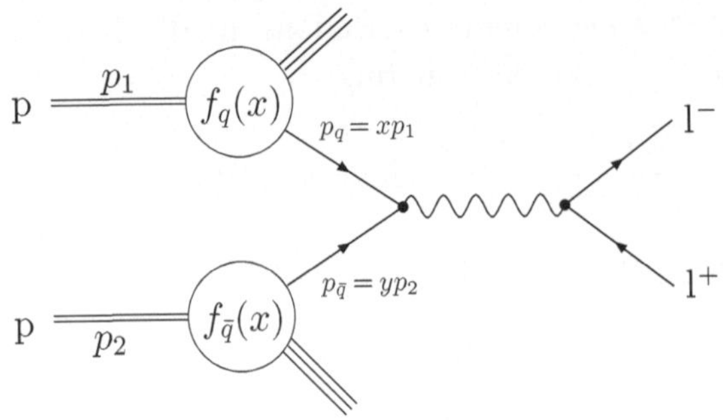

Figure 10.4. The Drell–Yan process.

Both reactions were analyzed in terms of the parton model, assuming that the constituents of protons are the quarks (Bjorken and Paschos, 1969). This will be further developed in the next two chapters, where the quark-parton content of hadrons becomes more evident.

10.5 The Drell–Yan process

The production of a massive photon or of a $W^{\pm}$ in hadron–hadron collisions and its subsequent decay has been successfully analyzed in terms of the parton model. The reactions (Drell and Yan, 1970)

$$\mathrm{p} + \mathrm{p} \to \gamma + \cdots \to \mu^+\mu^- + \mathrm{X}, \tag{10.53}$$

$$\bar{\mathrm{p}} + \mathrm{p} \to \mathrm{W} + \cdots \to \mathrm{e}^-\bar{\nu} + \mathrm{X} \tag{10.54}$$

are known as Drell–Yan processes. Together with deep inelastic scattering and electron–positron annihilation, these processes play an important role in determining the structure functions and in testing the parton model, including QCD corrections. The Drell–Yan process was especially important in formulating a strategy for seeking and discovering the W bosons.

To calculate the cross section corresponding to Fig. 10.4, we begin with the parton subprocess,

$$\sigma(\bar{\mathrm{q}}\mathrm{q} \to \ell^+\ell^-) = \frac{4\pi\alpha^2}{3Q^2} e_q^2. \tag{10.55}$$

In order to embed it in the hadronic process, we rewrite it as a differential cross section, $\mathrm{d}\sigma/\mathrm{d}Q^2$, for the production of a lepton pair with invariantmass $\sqrt{Q^2}$, where

$$Q^2 = \hat{s} = (p_q + p_{\bar{q}})^2, \tag{10.56}$$

$$\frac{\mathrm{d}\hat{\sigma}}{\mathrm{d}Q^2} = \frac{4\pi\alpha^2}{3Q^2}\, e_q^2 \delta(Q^2 - \hat{s}). \tag{10.57}$$

We envisage each hadron of momentum P being made up of partons carrying a longitudinal momentum xP. We make the idealization that the partons carry negligible transverse momentum. When the mass of the produced pair is very large, the cross section is the incoherent sum of the elementary subprocesses. In this case a quark of type q from one hadron annihilates with an antiquark of the same type from the other hadron. The probability of finding the quark with fractional momentum x is given by $f_q(x)$ and that for the antiquark by $f_{\bar{q}}(y)$. The hadronic cross section can now be obtained (Drell and Yan, 1970):

$$\frac{\mathrm{d}\sigma}{\mathrm{d}Q^2}(\mathrm{pp} \to \ell^+\ell^-\mathrm{X}) = \overline{\sum_q} e_q^2 \int \mathrm{d}x \int \mathrm{d}y\, f_q(x) f_{\bar{q}}(y) \frac{\mathrm{d}\hat{\sigma}}{\mathrm{d}Q^2}, \tag{10.58}$$

where the sum is over all possible $q\bar{q}$ pairs that can be formed from the constituents of the colliding protons and the average is over the number of initial $q\bar{q}$ states. This gives in the end an overall factor of $\frac{1}{3}$.

The q and $\bar{q}$ carry the fractions x and y of the proton momenta and the invariant mass becomes

$$\hat{s} = (xp_1 + yp_2)^2 \approx xys, \tag{10.59}$$

with $s \approx 2p_1 \cdot p_2$. The cross section now takes the form

$$\frac{\mathrm{d}\sigma}{\mathrm{d}Q^2} = \frac{4}{9}\frac{\pi\alpha^2}{Q^2} \int \mathrm{d}x \int \mathrm{d}y\, f_q(x) f_{\bar{q}}(y) \delta(Q^2 - xys). \tag{10.60}$$

After integration over y, we obtain the final result

$$\frac{\mathrm{d}\sigma}{\mathrm{d}Q^2} = \frac{4\pi\alpha^2}{9Q^2 s} \int_{Q^2/s}^{1} \frac{\mathrm{d}x}{x} f_q(x) f_{\bar{q}}\left(\frac{Q^2}{xs}\right). \tag{10.61}$$

To lowest order (without gluon emission) we expect a scaling result: the last integral depends on the ratio $\tau = Q^2/s$. The scaling is satisfied but the overall rate is modified by QCD corrections, which involve gluons. In this case the corrections are substantial and the reader should consult specialized articles for more details.

In addition to the quarks, the proton contains also gluons, i.e. vector mesons that mediate the strong interactions. This requires that for each hadron we must introduce a gluon distribution function: $g(x)$. In several processes gluons play an important role. For instance, the production of Higgses in high-energy colliders proceeds through the fusion of two gluons,

$$\mathrm{g} + \mathrm{g} \rightarrow \mathrm{H} \rightarrow \mathrm{ZZ}, \tag{10.62}$$

with g denoting gluons and the Higgs decaying to two lighter particles (in this case Z bosons). The hadronic reaction can be analyzed as a Drell–Yan process with the quarks of the intermediate states replaced by gluons.

Let us consider the process

$$\mathrm{p} + \mathrm{p} \rightarrow \mathrm{H} + \text{hadrons} \rightarrow \mathrm{Z\,Z} + \text{hadrons} \tag{10.63}$$

and denote by $\sigma_0(\mathrm{gg} \rightarrow \mathrm{H} \rightarrow \mathrm{ZZ})$ the point cross section for the production of two Z bosons. The gluon distribution function for protons has been measured in DESY experiments to be large at small values of x. The cross section for the production of Z pairs through two gluons with moments xp_1 and yp_2, respectively, is given by

$$\frac{\mathrm{d}\sigma_0}{\mathrm{d}Q^2} = \sigma_0(Q^2)\delta(Q^2 - xys). \tag{10.64}$$

The cross section for the proton–proton collision is

$$\frac{\mathrm{d}\sigma}{\mathrm{d}Q^2} = \int \frac{\sigma_0(Q^2)}{s} \frac{\mathrm{d}x}{x} g(x) g\left(\frac{Q^2}{xs}\right). \tag{10.65}$$

One usually takes the gluon structure functions from electron–proton-scattering experiments and extrapolates them to regions of small x and large Q^2 by means of the renormalization-group equations. In addition to the corrected structure functions, the calculations must include corrections to the gluon–Higgs-boson coupling induced again by virtual gluons.

Problems for Chapter 10

1. Show that Eq. (10.5) is determined up to an overall phase.
2. Determine the behaviour of $\varepsilon_\mu^{\mathrm{R,L}}$ under rotations around the z-axis. This can be done easily if you split $\varepsilon_\mu^{\mathrm{R,L}}$ into $\varepsilon_\mu^{(x)} = (0, 1, 0, 0)$ and $\varepsilon_\mu^{(y)} = (0, 0, 1, 0)$.
3. Prove Eq. (10.7) with the following *Ansatz*: $j_\mu^{\mathrm{lept}} = a\varepsilon_\mu^{\mathrm{S}} + b\varepsilon_\mu^{\mathrm{R}} + c\varepsilon_\mu^{\mathrm{L}}$. Determine a, b, and c using $k^\mu = (E, k_x, 0, k_z)$, $q^\mu = (\nu, 0, 0, q_z)$, and momentum conservation. Apply the high-energy limit $\nu \gg 2M$ and $Q^2 \ll \nu^2$ in order to obtain Eq. (10.7).
4. Derive Eq. (10.22) starting from Eq. (10.19).
5. In order to prove Eq. (10.30) rewrite j_μ^{lept} as

$$j_\mu^{\mathrm{lept}} = \frac{1}{2}\bar{u}'\gamma_\mu(1 + \gamma_5 + 1 - \gamma_5)u$$

and follow steps similar to those in the case of neutrino–hadron scattering.

6. Check the individual steps leading to Eq. (10.34).
7. Carry out the various steps leading to Eqs. (10.47) and (10.48).
8. Calculate the helicity cross section for a left-handed W scattered on quarks and show that it reproduces the result in Eq. (8.45).

References

Bjorken, J. D. (1969), *Phys. Rev.* **179**, 1547

Bjorken, J. D., and Paschos, E. A. (1969), *Phys. Rev.* **185**, 1975 (1970), *Phys. Rev.* **D1**, 3151

Drell, S. D., and Yan, T. M. (1970), *Phys. Rev. Lett.* **25**, 25

Feynman, R. P. (1969), *Phys. Rev. Lett.* **23**, 1415

11

Charged-current reactions

Charged-current interactions are the most frequent and occur in decays, as well as in particle reactions. They have been analyzed in many books, especially those written before 1970. Charged-current interactions, especially decays, were instrumental in establishing properties of the currents. We can classify them according to the degree of our theoretical understanding. The simplest reactions are purely leptonic. They are relatively simple to calculate, because the couplings of leptons to currents are precisely known and, now that the theory is renormalizable, we can include loop corrections. Some leptonic reactions were presented in Chapter 8. We shall not study them further.

The next class of reactions consists of the semileptonic ones, which can also be treated successfully with various theoretical methods. They involve a single coupling of the currents to hadrons, which can be understood at low energy and/or at low momentum transfer in terms of form factors. They are also understood at high energies in terms of the short-distance behavior of the currents. We shall study several processes in this chapter: deep inelastic scattering and quasi-elastic scattering.

Non-leptonic interactions are the most difficult to analyze. They do not include any leptons and involve both strong and weak interactions. The interplay between the two interactions is still a developing field of research.

11.1 Deep inelastic scattering

High-energy neutrino interactions have been used to probe the inner structure of protons and neutrons: these studies were crucial for establishing the quark substructure of matter and giving quantitative support to the field theory of quark interactions (quantum chromodynamics). In Chapter 10 we described the general structure of the cross sections and some consequences of the scaling phenomenon.

Then we showed that the general features can be explained in terms of the quark–parton model. Many more properties and correlations with other reactions have been understood and we discuss them here in greater detail.

11.1.1 Scaling and the charge of the quarks

The electroproduction reactions couple to the charge of the quarks, in contrast to the neutrino reactions, which couple to the weak isospin. Comparison of the two processes gives an indication regarding the charge of the constituents.

The electroproduction cross section is

$$\begin{aligned}\frac{d\sigma}{dQ^2\,d\nu} &= \frac{4\pi\alpha^2}{Q^4}\int dx\,\delta\left(\nu - \frac{Q^2}{2Mx}\right)\sum_i e_i^2[q_i(x)+\bar{q}_i(x)] \\ &= \frac{4\pi\alpha^2}{Q^4}\frac{x}{\nu}\sum_i e_i^2[q_i(x)+\bar{q}_i(x)], \qquad (11.1)\end{aligned}$$

with the normalization chosen to reproduce the Mott cross section. We obtain the structure function

$$F_2^{\mathrm{ep}}(x) = x\left\{\frac{4}{9}[u(x)+\bar{u}(x)] + \frac{1}{9}\big[d(x)+\bar{d}(x)+s(x)+\bar{s}(x)\big]\right\}, \qquad (11.2)$$

where $u(x)$, $d(x)$, $\bar{u}(x)$, and $\bar{d}(x)$ are the quark distribution functions in the proton. Similar equations hold for electron–neutron scattering, in which the exchanges $\mathrm{u} \leftrightarrow \mathrm{d}$ and $\bar{\mathrm{u}} \leftrightarrow \bar{\mathrm{d}}$ take place. The structure function on an isoscalar target is the average over protons and neutrons,

$$F_2^{\mathrm{eN}}(x) = x\left\{\frac{5}{18}\big[u(x)+\bar{u}(x)+d(x)+\bar{d}(x)\big] + \frac{1}{9}[s(x)+\bar{s}(x)]\right\}, \qquad (11.3)$$

where the factor of $5/18$ follows directly from the fractional charges of the quarks.

For neutrino-induced reactions the structure functions are expressed in terms of the quark distributions

$$F_2^{\nu\mathrm{p}}(x) = 2x[q_i(x)+\bar{q}_i(x)], \qquad (11.4)$$

$$F_2^{\nu\mathrm{p}}(x) = 2xF_1^{\nu\mathrm{p}}(x) \qquad \text{(Callan and Gross, 1969)}, \qquad (11.5)$$

$$xF_3^{\nu\mathrm{p}}(x) = 2x[q_i(x)-\bar{q}_i(x)]. \qquad (11.6)$$

For neutrinos the following elementary processes are possible:

$$\nu\mathrm{d} \to \mu^-\mathrm{u},$$

$$\nu\bar{\mathrm{u}} \to \mu^-\bar{\mathrm{d}};$$

and for antineutrinos

$$\bar{\nu}\mathrm{u} \to \mu^+\mathrm{d},$$
$$\bar{\nu}\bar{\mathrm{d}} \to \mu^+\bar{\mathrm{u}}.$$

We shall assume in this section that the Cabbibo angle is zero, so that scatterings from strange quarks are neglected. Then we obtain the structure functions

$$\begin{aligned} F_2^{\nu\mathrm{p}}(x) &= 2x[d(x) + \bar{u}(x)], \\ F_2^{\bar{\nu}\mathrm{p}}(x) &= 2x\left[\bar{d}(u) + u(x)\right] = f_2^{\nu\mathrm{n}}(x) \end{aligned} \tag{11.7}$$

and, for isoscalar targets,

$$F_2^{\nu\mathrm{N}}(x) = x\left[u(x) + d(x) + \bar{u}(x) + \bar{d}(x)\right] \tag{11.8}$$

and

$$x F_3^{\nu\mathrm{N}}(x) = x\left[u(x) + d(x) - \bar{u}(x) - \bar{d}(x)\right]. \tag{11.9}$$

Since the strange-quark structure functions $s(x)$ and $\bar{s}(x)$ are relatively small, we can neglect them and obtain from Eqs. (11.3) and (11.8) the ratio

$$\frac{F_2^{\mathrm{eN}}(x)}{F_2^{\nu\mathrm{N}}(x)} = \frac{5}{18}. \tag{11.10}$$

The ratio measures the average charge of the quarks and it indicates that the charges of the constituents are fractional. These and other relations have shown that the constituents of hadrons which couple in deep inelastic scattering carry the quantum numbers of the quarks.

11.1.2 Spin of the quarks

In the V–A theory all fermions participate in the weak interactions as left-handed particles and all antifermions as right-handed particles. For antineutrino–quark scattering, helicity conservation requires that the process in the center-of-mass frame vanishes at $\theta_{\mathrm{cm}} = 180^\circ$ as shown in Fig. 11.1.

The cross section has the angular dependence

$$\frac{\mathrm{d}\sigma^{\bar{\nu}\mathrm{q}}}{\mathrm{d}\cos\theta_{\,\mathrm{cm}}} \propto (1 + \cos\theta_{\mathrm{cm}})^2. \tag{11.11}$$

The center-of-mass angle θ_{cm} is related to the laboratory energies by

$$\frac{1 + \cos\theta_{\mathrm{cm}}}{2} = \frac{E_\mu}{E_\nu} = 1 - y, \tag{11.12}$$

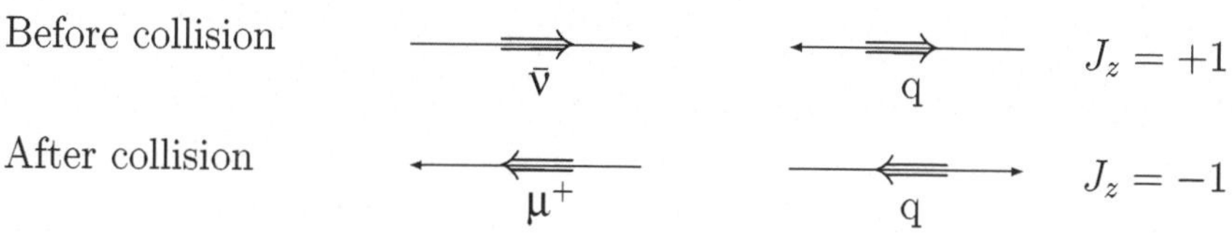

Figure 11.1. Production of a forbidden configuration by helicity conservation in antineutrino–quark scattering.

Table 11.1. *Angular dependences of the reactions*

Process	J_z	y Dependence
$\nu q,\ \bar{\nu}\bar{q} : \underset{\nu}{\Leftarrow}\ \underset{q}{\Rightarrow}$	0	1
$\nu\bar{q},\ \bar{\nu}q : \underset{\nu}{\Leftarrow}\ \underset{\bar{q}}{\Leftarrow}$	1	$(1-y)^2$

which is easily obtained by evaluating the ratio $k' \cdot p/(k \cdot p)$ in the center-of-mass system. In this pictorial manner we understand the y dependence of the cross section:

$$\frac{d\sigma^{\bar{\nu}q}}{dy} \propto (1-y)^2. \tag{11.13}$$

A similar study of the reaction

$$\nu + d \to \mu^- + u \tag{11.14}$$

shows that there is no reason for the cross section to vanish in any direction. In this way we construct Table 11.1.

The cross sections in terms of the various species attain the form with $q(x)$ and $\bar{q}(x)$ contributions from spin-$\frac{1}{2}$ constituents and $k(x)$ distributions for spin-zero constituents:

$$\frac{d\sigma^{\nu N}}{dx\,dy} = \frac{G^2 M E_\nu}{\pi} x\big[q(x) + (1-y)^2\bar{q}(x) + (1-y)k(x)\big], \tag{11.15}$$

$$\frac{d\sigma^{\bar{\nu}N}}{dx\,dy} = \frac{G^2 M E_{\bar{\nu}}}{\pi} x\big[(1-y)^2 q(x) + \bar{q}(x) + (1-y)k(x)\big]. \tag{11.16}$$

The experimental results indicate that the scattering occurs on spin-$\frac{1}{2}$ constituents and that the content of scalar constituents is very small. The y distribution for antineutrinos is not exactly zero at $y = 1$ because protons and neutrons contain a sea of quark–antiquark pairs in addition to their valence quarks. These pairs are created by the emission of vector particles, the gluons, which also bind the quarks into hadrons.

11.1.3 Sum rules

In the quark parton model, the distribution functions indicate how the quantum numbers are distributed within hadrons. Thus integrals of distribution functions must reproduce the quantum numbers of the target. For instance, $\int_0^1 s(x)\mathrm{d}x$ gives the probability of finding a strange quark, with any momentum, within a proton. Since the proton has zero strangeness,

$$\int_0^1 [s(x) - \bar{s}(x)]\mathrm{d}x = 0. \tag{11.17}$$

Similarly we can compute the baryon number and isospin of a proton:

$$\frac{1}{3}\int_0^1 \left[u + d - \bar{u} - \bar{d}\,\right]\mathrm{d}x = 1 \quad \text{(baryon)}, \tag{11.18}$$

$$\int_0^1 \left[(u - d) - (\bar{u} - \bar{d})\right]\mathrm{d}x = 1 \quad \text{(isospin)}. \tag{11.19}$$

Such relations are known as sum rules and have been determined by combining data from various processes.

A good example is the Adler (1965) sum rule

$$S_{\mathrm{A}} = \frac{1}{2}\int \left[F_2^{\bar{\nu}\mathrm{p}}(x) - F_2^{\nu\mathrm{p}}(x)\right]\frac{\mathrm{d}x}{x} = 1, \tag{11.20}$$

which follows from the isospin relation in (11.19). The Adler sum rule follows from current algebra and it must be valid for each value of Q^2. In fact, it is a consequence of the commutator of two isospin charges and as such is very reliable. Experimental results give the value

$$S_{\mathrm{A}} = 1.08 \pm 0.20, \tag{11.21}$$

which is in good agreement but the error is relatively large.

A fast convergent sum rule is the Gross–Llewellyn Smith (1969) sum rule

$$\begin{aligned}\int_0^1 \left[F_3^{\nu\mathrm{p}}(x, Q^2) + F_3^{\nu\mathrm{n}}(x, Q^2)\right]\mathrm{d}x &= 6\left[1 - \frac{\alpha_{\mathrm{s}}(Q^2)}{\pi}\right] \\ &= 5.4 \quad \text{at} \quad Q^2 = 3\ \mathrm{GeV}^2 \\ &= 5.00 \pm 0.16 \quad \text{(experiment)}. \end{aligned} \tag{11.22}$$

The right-hand side includes first-order QCD corrections. Again the agreement is very good.

Finally, the integral $\int_0^1 xq(x)\mathrm{d}x$ gives the fraction of the proton's momentum carried by the q quark. Thus

$$\sum = \int_0^1 x\big[u(x) + d(x) + s(x) + \bar{u}(x) + \bar{d}(x) + \bar{s}(x)\big]\mathrm{d}x = 0.54 \quad (11.23)$$

is the momentum carried by all the quarks inside the proton. This integral was determined by combining data from several processes. It is much less than unity, indicating that the quarks carry one half of the proton's momentum. The remaining half must be carried by other particles, which do not interact directly with the currents. They are the gluons of quantum chromodynamics.

11.2 Evolution of distribution functions

The algebraic relations discussed in this and the previous chapter assumed point-like constituents within the nucleon. We know from previous advances in physics that a particle that looks point-like on one resolution scale reveals substructure at a higher resolution. The scaling phenomenon and the numerous quantum-number relations revealed a point-like structure, but deviations from scaling indicate the existence of additional structure. In fact, it has been established that the variation of the structure functions with Q^2 is due to the emission of vector particles: the gluons. As was mentioned earlier, they carry the other half of the momentum of the nucleons, that was missing in the momentum sum rule.

The additional structure is introduced by the theory of strong interactions known as quantum chromodynamics (QCD). There are many indications that each flavor of quarks comes in three colors: red, white, and blue. The names for the colors are arbitrary, but the fact that there are three is important. The quarks interact with each other by the exchange of vector bosons that change the color of the quarks.

The theory of the strong interaction – QCD – is a non-Abelian gauge theory based on the group $\mathrm{SU}(3)_\mathrm{c}$. Each quark species – up, down, strange, ...– forms a triplet in color space and has a coupling g_s to the eight gluons, G_μ^α, which belong to the adjoint representation of SU(3) color. We write the color triplet as

$$q(x) = \begin{pmatrix} q_\mathrm{r} \\ q_\mathrm{w} \\ q_\mathrm{b} \end{pmatrix}$$

and the Lagrangian

$$\mathcal{L}_\mathrm{QCD} = \bar{q}(x)\mathrm{i}\gamma^\mu\left[\partial_\mu + \frac{\mathrm{i}}{2}g_\mathrm{s}\lambda^\alpha G_\mu^\alpha(x)\right]q(x), \quad (11.24)$$

with λ^α the Gell-Mann matrices of $SU(3)_c$ and $G^\alpha_\mu(x)$ with $\alpha = 1, 2, \ldots, 8$ the eight gluons. There is no mass term for the gluons that leaves the color symmetry exact. QCD makes dramatic predictions. The first one concerns the coupling constant.

In field theories coupling constants and other observables are modified by higher-order corrections that involve loops. Many loop diagrams are divergent, which demands special handling of them. When all the infinities from loop diagrams are absorbed into the definition of couplings, masses, and other parameters of the original Lagrangian, we say that the theory is renormalizable. In these theories we can calculate physical observables with high precision. QCD and the electroweak theory are renormalizable. It is beyond the scope of this book to describe or prove renormalization. Instead, we shall describe a few cases of higher-order corrections in order to demonstrate the methods entering these calculations. Furthermore, we describe some properties of field theories that have significant impact on properties of weak interactions.

One quantity modified by loop corrections is the strong coupling constant g_S. The infinities introduced by higher orders are absorbed into the redefinition of the coupling constant. Since the corrections involve the addition of infinite quantities, the numerical value of the coupling is unknown and must be determined experimentally. Thus α_s is measured at a specific reference scale μ_0, known as the renormalization point. In many cases the reference scale μ is identified with the momentum flowing through the vertex. The arbitrariness of the reference point leads to a differential equation – the renormalization-group equation. To be specific, the change of $\alpha(\mu) = g_s^2/(4\pi)$ with respect to the reference point μ satisfies the equation

$$\mu \frac{d\alpha}{d\mu} = \beta(\alpha), \tag{11.25}$$

where $\beta(\alpha)$ represents the sum of higher-order corrections and is of the form

$$\beta(\alpha) = \beta_0\alpha^3 + O(\alpha^5). \tag{11.26}$$

The constant β_0 and higher terms are determined in perturbation theory. The solution is obtained as

$$\int_{\alpha(\mu_0)}^{\alpha(\mu)} \frac{d\alpha}{\beta(\alpha)} = \ln\left(\frac{\mu}{\mu_0}\right) \tag{11.27}$$

or, keeping the leading term on the right-hand side of Eq. (11.26), we obtain

$$\alpha(\mu) = \frac{\alpha(\mu_0)}{1 + \beta_0\alpha(\mu_0)\ln\left(\mu^2/\mu_0^2\right)}\,. \tag{11.28}$$

This states that, knowing the coupling constant at the reference scale μ_0, we can predict its value at another scale μ. The coupling constant is no longer a constant but runs with momentum; hence the name running coupling constant.

An important property of QCD is that the value of $\beta_0 = 11 - \frac{2}{3}N_\mathrm{f}$, with N_f the number of generations, is positive. As the momentum increases, the coupling constant decreases and, at very high momentum, $\alpha_\mathrm{s}(p)$ is so small that perturbation theory is applicable. This provides a justification of scaling and of the parton model. It also goes beyond scaling, by predicting modifications introduced by the emission of gluons. The corrections are functions of Q^2 producing predictable violations of scaling. The corrections have been studied extensively in perturbation theory and compared with many experimental results.

There is an extensive list of articles in which violations of scaling have been computed and are discussed in detail. Experimentally, the changes have been observed with the structure functions increasing for small x as functions of Q^2 and decreasing for $x > 0.4$.

The second prediction concerns the production of gluons, which are emitted by the accelerating particles. The quarks produced materialize into hadrons and produce jets of particles. Similarly, the gluons also produce jets of hadrons. Consequently we expect some reactions to produce two jets (from $\mathrm{q\bar{q}}$ pairs) and others three jets (from $\mathrm{q\bar{q}g}$ production). Three-jet events have been observed in electron–positron-annihilation reactions. The production of gluons implies that they also exist within hadrons and are responsible for the missing momentum in the sum rule in Eq. (11.23). Consequently the experimental results have been analyzed with the inclusion of an additional distribution function for the gluons. A dramatic property of the gluon distribution function is its rapid increase at small x.

For small momenta the coupling constant grows and becomes very big, making a perturbative description impossible. It is customary to denote by Λ^2 the scale of Q^2 at which the denominator becomes zero. This happens at

$$\Lambda^2 = \mu_0^2 \mathrm{e}^{-1/[\beta_0 \alpha(\mu_0)]}. \tag{11.29}$$

It follows now that the coupling constant can be rewritten

$$\alpha(\mu) = \frac{1}{\beta_0 \ln(\mu^2/\Lambda^2)}. \tag{11.30}$$

We can think of Λ as the boundary between the region where quarks and gluons appear as quasi-free particles and the world of bound states like protons, pions, etc. At momenta smaller than Λ, the strong interaction becomes so strong that the quarks cannot come out as free particles, but remain confined within hadrons. This property of confined quarks has not been proved yet, and thus it is difficult to judge which among several approaches may ultimately be the most productive.

It is evident that QCD and its consequences form an extensive and exciting topic, which is, however, beyond the scope of this book. We shall have occasion to return to QCD in Section 15.6, where we discuss the effective Hamiltonian for low-energy weak interactions. As a last topic concerning the charged-current interactions we discuss in the next section quasi-elastic scattering.

11.3 Quasi-elastic scattering

In contrast to deep inelastic scattering, quasi-elastic scattering gives information on the static properties of the proton and the neutron. In fact, the first experiments with neutrino beams measured the reactions shown in Fig. 11.2:

$$\nu(k) + \mathrm{n}(p) \to \mu^-(k') + \mathrm{p}(p'), \tag{11.31}$$

$$\bar{\nu}(k) + \mathrm{p}(p) \to \mu^+(k') + \mathrm{n}(p'), \tag{11.32}$$

which are still interesting on several accounts. For instance, we would like to determine their form factors accurately and check their relation to the electromagnetic form factors. Furthermore, the quasi-elastic cross sections reach constant values for neutrino energies greater than 2.0 GeV. This property has been used for measuring the flux of neutrino beams and is still useful. Low-energy neutrino interactions recently started being used efficiently for studying neutrino oscillations. For all these reasons we present in this chapter an explicit calculation.

For low energies relative to the mass of the W boson, interactions of neutrinos can be written as a leptonic current times a hadronic current:

$$\mathcal{M} = \frac{G}{\sqrt{2}}\bar{u}(k')\gamma_\mu(1-\gamma_5)u(k)\langle p|J_\mu^+|n\rangle.$$

The hadronic current has a complicated structure produced by the motion of the quarks within hadrons. We define the vector form factors of the charged current as

$$\langle p|V_\mu^+|n\rangle = \bar{u}(p')\left(\gamma_\mu F_1^+ + \mathrm{i}\frac{\sigma_{\mu\nu}q^\nu}{2M}F_2^+ + \frac{q_\mu}{M}F_3^+\right)u(p). \tag{11.33}$$

Denoting, as before, the isovector form factor of the electromagnetic current by F_1^{V}, we obtain

$$F_1^+(q^2) = -2F_1^{\mathrm{V}}(q^2). \tag{11.34}$$

The factor 2 comes from the normalization of $V_\mu^+ = V_\mu^1 + \mathrm{i}V_\mu^2$ and the Clebsch–Gordan coefficients, which are $\sqrt{\frac{2}{3}}$ for the charged current and $-\sqrt{\frac{1}{3}}$ for

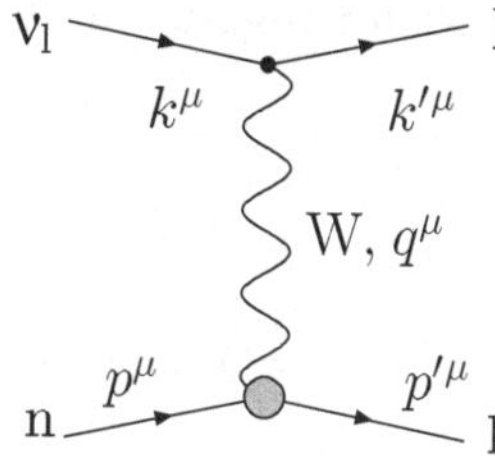

Figure 11.2. Quasi-elastic scattering.

the electromagnetic current. For the following calculation it is convenient to write the hadronic current in a modified form:

$$\langle p|J_\mu^+|n\rangle = \bar{u}(p')\bigg[g_V\gamma_\mu + f_V\frac{(p+p')_\mu}{2M} + h_V\frac{q_\mu}{2M} + g_A\gamma_\mu\gamma_5 + f_A i\sigma_{\mu\nu}\frac{q^\nu\gamma_5}{2M} + h_A\frac{q_\mu\gamma_5}{2M}\bigg]u(p), \qquad (11.35)$$

where we left out the factor $\cos\theta_c$ arising from the Cabibbo angle. In reducing (11.33) to (11.35) we used the Gordon decomposition formula, which gives the relations $g_V = F_1^+ + F_2^+$, $f_V = -F_2^+$, and F_3^+ vanishes as explained in Chapter 2. The term $h_A q_\mu \gamma_5$ contributes to the cross section terms proportional to the lepton masses and will be omitted. In Chapters 1 and 2 we discussed the fact that charge conjugation and time-reversal together require h_V and f_A to vanish. On eliminating these three form factors, we obtain a simplified form for the matrix element which we shall use in this section:

$$\langle p|J_\mu^+|n\rangle = \bar{u}(p')\bigg[g_V\gamma_\mu + f_V\frac{(p+p')_\mu}{2M} + g_A\gamma_\mu\gamma_5\bigg]u(p). \qquad (11.36)$$

The electromagnetic form factors are known from electron-scattering experiments on protons and neutrons. Similar values of the axial form factor at low values of Q^2 have been measured in β-decay. We shall use this information at the end of this section.

The calculation of the cross section is now straightforward but tedious. Since the calculation of quasi-elastic scattering is not easily available in books, I give a few intermediate steps. There is a second reason: the elastic scattering for neutral currents has a similar functional form that is obtained by replacing the form factors by those of neutral currents. Studies of quasi-elastic scattering frequently use a formula in terms of Mandelstam variables (Llewellyn Smith, 1974). We derive here two more formulas that are convenient for taking limits in specific kinematic regions.

The kinematics for the process are simplest in the laboratory frame where the nucleon is at rest:

$$
\begin{aligned}
p \cdot k &= p' \cdot k' = ME, \\
p \cdot k' &= p' \cdot k = ME + \frac{q^2}{2}, \\
k \cdot k' &= \frac{m_\mu^2}{2} - \frac{q^2}{2}, \\
p \cdot p' &= \frac{M^2}{x} - \frac{q^2}{2}, \\
Q^2 &= 2M\nu = 4EE' \sin^2\left(\frac{\theta}{2}\right).
\end{aligned}
\tag{11.37}
$$

The square of the hadronic tensor is obtained from (11.36)

$$
\begin{aligned}
H_{\mu\nu} &= \left(g_V^2 + g_A^2\right)\left(p_\mu p'_\nu + p_\nu p'_\mu - g_{\mu\nu} p \cdot p'\right) - 2\mathrm{i} g_V g_A \epsilon_{\mu\nu\gamma\delta} p^\gamma p'^\delta \\
&\quad + M^2\left(g_V^2 - g_A^2\right) g_{\mu\nu} + \left(f_V^2 \frac{p \cdot p' + M^2}{4M^2} + f_V g_V\right)\left(p_\mu + p'_\mu\right)\left(p_\nu + p'_\nu\right).
\end{aligned}
\tag{11.38}
$$

The first line of this equation follows from Eq. (8.37) and the remaining ones from a straightforward calculation. The similarities between Section 8.3 and the present one can be used for comparisons. For instance, the inner product of the leptonic tensor in Eq. (8.36) with $H_{\mu\nu}$ leads to a matrix element that can be expressed in terms of the scattering angle. From the matrix element and the phase-space integral we arrive at the differential cross section

$$
\begin{aligned}
\frac{\mathrm{d}\sigma}{\mathrm{d}E'} &= \frac{G_F^2}{2\pi} M \frac{E'}{E} \left\{ \left(g_V^2 + g_A^2\right)\left[1 + \frac{Q^2}{2M^2}\sin^2\left(\frac{\theta}{2}\right)\right] + \left(g_A^2 - g_V^2\right)\sin^2\left(\frac{\theta}{2}\right) \right. \\
&\quad \left. - 2 g_A g_V \left(\frac{E + E'}{M}\right)\sin^2\left(\frac{\theta}{2}\right) + \left[f_V^2\left(1 + \frac{Q^2}{4M^2}\right) + 2 f_V g_V\right]\cos^2\left(\frac{\theta}{2}\right) \right\}.
\end{aligned}
\tag{11.39}
$$

Several limiting cases are now interesting. For $g_A = 0$ the cross section depends only on the vector terms and the functional form agrees with the Rosenbluth formula. Differences between electroproduction and neutrino-induced formulas arise from the photon propagator and the coupling constants.

Alternatively, we may combine the g_V and g_A terms and obtain another expression for the cross section:

$$\frac{d\sigma}{dE'} = \frac{G_F^2 M}{4\pi}\left\{(g_V - g_A)^2 + (g_V + g_A)^2\left(\frac{E'}{E}\right)^2 + \left(g_A^2 - g_V^2\right)\frac{M\nu}{E^2}\right.$$
$$\left. + \frac{1}{2}\left[f_V^2\left(1+\frac{Q^2}{4M^2}\right) + 2f_V g_V\right]\left[\left(1+\frac{E'}{E}\right)^2 - \frac{Q^2}{E^2}\left(1+\frac{Q^2}{4M^2}\right)\right]\right\}. \tag{11.40}$$

For $f_V = 0$ the interaction of the neutrino has the same functional form as neutrino–electron scattering and the expression above agrees with Eq. (8.43).

The two equations for quasi-elastic scattering presented already are convenient for taking specific limits. It is customary, however, to use another formula, which expresses the differential cross section in terms of Mandelstam variables s, u, $t = q^2$ (Llewellyn Smith 1974). Most recent analyses use it and have been able to account for the experimental data in terms of three form factors. The vector form factors are related to those measured in electromagnetic reactions. The axial form factor is parametrized

$$g_A(q^2) = \frac{g_A(0)}{\left(1 - q^2/M_A^2\right)^2}, \tag{11.41}$$

with $g_A(0) = 1.26$ and $M_A = 1.00 \pm 0.05\ \text{GeV}/c^2$ (its precise value is still being debated among the experts).

References

Adler, S. L. (1965), *Phys. Rev.* **143**, 1144
Callan, C. G., and Gross, D. (1969), *Phys. Rev. Lett.* **22**, 156
Gross, D., and Llewellyn Smith, C. H. (1969), *Nucl. Phys.* **B14**, 337
Llewellyn Smith, C. H. (1974), *Phys. Rep.* **3C**, 264

Select bibliography

Bjorken, J. D. and Paschos, E. A. (1970), *Phys. Rev.* **D1**, 3151
Feynman, R. P. (1972), *Photon–Hadron Interactions* (Reading, MA, W. A. Benjamin)

12

Neutral currents in semileptonic reactions

12.1 Neutrino–hadron neutral-current interactions

The first experimental support for the electroweak theory came from the observation of neutral currents in semileptonic reactions. Neutral currents appear because the product SU(2) × U(1) contains two neutral generators. We have shown that one linear superposition of generators is the electromagnetic current and the second is a neutral current. In Chapter 8 we discussed leptonic neutral-current reactions. In this chapter we deal with the observation of neutral currents in semileptonic reactions and, in particular, neutrino–hadron interactions.

The coupling of the Z_μ boson to leptons was given in Eq. (8.11) and that to quarks in Eq. (9.20). The neutral-current neutrino–hadron interactions have the general form

$$\begin{aligned} H_{\text{eff}} &= \frac{G}{\sqrt{2}}\left[\bar{\nu}\gamma^\mu(1-\gamma_5)\nu\right]\sum_{i=1}^{3}\left(\bar{q}_i\,\tau_3\gamma_\mu q_i - \sin^2\theta_{\mathrm{W}}\,\bar{q}_i\,Q\gamma_\mu q_i\right) \\ &= \frac{G}{\sqrt{2}}\left[\bar{\nu}\gamma^\mu(1-\gamma_5)\nu\right]\left(xV_\mu^3 + yA_\mu^3 + \gamma V_\mu^0 + \delta A_\mu^0\right), \end{aligned} \tag{12.1}$$

where V_μ^3 and A_μ^3 are the isospin partners of the charged currents. V_μ^0 and A_μ^0 are isoscalar currents for which there are several possibilities. For comparison we give the normalization of V_μ^3 in terms of quarks:

$$\begin{aligned} V_\mu^3 &= \frac{1}{2}\left(\bar{u}\gamma_\mu u - \bar{d}\gamma_\mu d\right), \\ A_\mu^3 &= \frac{1}{2}\left(\bar{u}\gamma_\mu\gamma_5 u - \bar{d}\gamma_\mu\gamma_5 d\right). \end{aligned} \tag{12.2}$$

We could have defined V_μ^3 abstractly, in terms of its isospin transformation properties, but, now that quarks permeate our daily language, this notation is appropriate. The interested reader can always revert to the transformation properties. Similarly, we define the isoscalar currents

$$
\begin{aligned}
V_\mu^0 &= \frac{1}{2}(\bar{u}\,\gamma_\mu u + \bar{d}\gamma_\mu\, d) + \cdots, \\
A_\mu^0 &= \frac{1}{2}(\bar{u}\,\gamma_\mu\gamma_5 u + \bar{d}\gamma_\mu\gamma_5\, d) + \cdots,
\end{aligned}
\tag{12.3}
$$

where $\cdots$ involve $\bar{s}s$ and $\bar{c}c$ terms. With this normalization, the isoscalar piece of the electromagnetic current is $\frac{1}{3}V_\mu^0$. In the electroweak theory

$$
\begin{aligned}
x &= 1 - 2\sin^2\theta_{\mathrm{W}}, \qquad & y &= -1, \\
\gamma &= -\frac{2}{3}\sin^2\theta_{\mathrm{W}}, \qquad & \delta &= 0.
\end{aligned}
\tag{12.4}
$$

The vanishing of δ is a specific property of the standard model when we consider only up and down quarks. It is non-zero as soon as strange and heavier quarks or higher-order corrections are introduced.

The weak mixing angle θ_{W} is the same angle as that introduced in the leptonic sector. The first issue was the existence of the neutral currents. This was a difficult experimental problem because neutral-current interactions were new and the experiments had a large neutron background.

After the discovery of neutral currents, there was still interest in establishing that they belonged to the standard model. As problems, there remained

(i) verification of the Lorentz structure of neutral currents as vector and axial-vector operators, and
(ii) verification of the internal symmetry structure as a superposition of isovector and isoscalar operators in terms of a mixing parameter: $\sin^2\theta_{\mathrm{W}}$.

In analyzing these issues there are two separate kinematic regions where we know the hadronic matrix elements of the currents. One region is deep inelastic scattering, where the structure functions have been measured and have been explained successfully in terms of quark-parton distribution functions. The other region involves low-energy experiments, for which form factors for elastic scattering and the excitation of the $\Delta(1232)$ resonance are already known. In the next few sections we study reactions that allow us to decipher the couplings of neutral currents to hadrons.

When the standard model became popular, it appeared very important to discover neutral currents. It was also fortunate that experiments with the capability of searching for them were running or were beginning to run. It was not clear, however, how large neutral-current cross sections should be. There was a need for theoretical predictions. At that time the quark-parton model was in its infancy and its predictions were frequently questioned.

Thus theoretical predictions were carried out at two levels. One approach was through the symmetry properties of the currents relating V_μ^3 and V_μ^0 to the charged and electromagnetic currents. The other approach was to calculate cross sections in the quark-parton model. Nowadays we know that both approaches are correct.

12.2 Model-independent predictions

The simplest processes to consider are those involving total cross sections on isospin neutral targets. We define

$$\sigma_- = \frac{1}{2}\left[\sigma\,(\nu + \mathrm{p} \to \mu^- + \mathrm{X}_1) + \sigma(\nu + \mathrm{n} \to \mu^- + \mathrm{X}_2)\right]$$

and

$$\sigma_0 = \frac{1}{2}\left[\sigma(\nu + \mathrm{p} \to \nu + \mathrm{X}_3) + \sigma(\nu + \mathrm{n} \to \nu + \mathrm{X}_4)\right]. \tag{12.5}$$

An incoherent sum over all possible final states that yields an isoscalar final state is assumed. For the charged-current cross section we write

$$\sigma_- = V + A + I, \tag{12.6}$$

where V comes from the vector current alone, A from the axial current alone, and I is the interference term. We can represent them as follows:

$$V = \sum |\langle X|\,\varepsilon \cdot V\,|p\rangle|^2, \qquad A = \sum_{x,\varepsilon} |\langle X|\,\varepsilon \cdot A|p\rangle|^2,$$

and

$$I = 2\sum \mathrm{Re}(\langle X|\varepsilon \cdot V\,|p\rangle^* \langle X|\,\varepsilon \cdot A\,|p\rangle), \tag{12.7}$$

with the sums running over all final states and polarizations of the W boson. In Eqs. (12.6) and (12.7) an average over protons and neutrons is understood.

The vector currents are isovector quantities that are related to the isovector part of the neutral current through an isospin rotation. The neutral current contains in addition an isoscalar term, but, since we consider isoscalar target and isoscalar final states, the isoscalar–isovector interference drops out. It follows now that

$$\sigma_0 = \frac{1}{2}(x^2 V + xI + A + y^2 S), \tag{12.8}$$

where S is the contribution of the isoscalar current. The overall factor of $1/2$ follows from the fact that the charged current transforms like the generator $\sqrt{2}\tau^+$ of SU(2) and the neutral current like τ^3. Since $y^2 \geq 0$,

$$R = \frac{\sigma_0}{\sigma_-} \geq \frac{1}{2}\frac{A + xI + x^2 V}{A + I + V}. \tag{12.9}$$

Furthermore, Schwarz's inequality implies

$$4AV \geq I^2. \tag{12.10}$$

On combining the two inequalities (see Problem 1), we arrive at

$$R \geq \frac{1}{2}\left[1 - (1-x)\left(\frac{V}{A + I + V}\right)^{\frac{1}{2}}\right]^2. \tag{12.11}$$

The term V can be deduced from knowledge of the isovector contribution to the electroproduction cross section

$$\sigma_{\mathrm{em}} = \frac{1}{2}\big[\sigma(\mathrm{e} + \mathrm{p} \to \mathrm{e} + \mathrm{x}_1) + \sigma(\mathrm{e} + \mathrm{n} \to \mathrm{e} + \mathrm{x}_2)\big]. \tag{12.12}$$

Not knowing the isoscalar contribution, we use again inequalities,

$$V \leq \frac{G}{\pi}\frac{Q^4}{4\pi\alpha^2}\sigma_{\mathrm{em}} = V_{\mathrm{em}}, \tag{12.13}$$

which gives the final result

$$R \geq \frac{1}{2}\left[1 - 2\sin^2\theta_{\mathrm{W}}\left(\frac{V_{\mathrm{em}}}{\sigma_-}\right)^{\frac{1}{2}}\right]^2. \tag{12.14}$$

This derivation makes judicious use of inequalities. Within the electroweak theory the method is model-independent and holds for many physical processes. When we plot R versus $\sin^2\theta_{\mathrm{W}}$ there is a minimum for the ratio; for similar inequalities see Pais and Treiman (1972).

One may also use reactions induced by antineutrinos to obtain additional relations. On going over to antineutrinos one must change the sign of the interference term I. The charged- and neutral-current cross sections on isoscalar targets are, respectively,

$$\sigma_+ = (V + A - I), \tag{12.15}$$

$$\bar{\sigma}_0 = \frac{1}{2}(A + x^2 V - xI + y^2 S). \tag{12.16}$$

On combining Eqs. (12.6), (12.8), (12.15), and (12.16), we obtain (Paschos and Wolfenstein, 1973)

$$R_- = \frac{\sigma_0 - \bar{\sigma}_0}{\sigma_- - \sigma_+} = \frac{1}{2}(1 - 2\sin^2\theta_W), \tag{12.17}$$

$$R_+ = \frac{\sigma_0 + \bar{\sigma}_0}{\sigma_- + \sigma_+} = \left(\frac{1}{2} - \sin^2\theta_W + \frac{10}{9}\sin^4\theta_W\right). \tag{12.18}$$

These relations are truly independent of any details of scaling violations and eliminate some theoretical corrections inherent in the quark-parton method. They are frequently used to determine the mixing angle $\sin^2\theta_W$.

In the above derivations we set the parameter

$$\rho = \frac{M_W^2}{M_Z^2\cos^2\theta_W} \tag{12.19}$$

equal to unity. This is the lowest-order value, which appears also in Eq. (8.19), but radiative corrections will modify it. Extensive analyses of the data including radiative corrections from the top quark and the Higgs meson gave the value

$$\rho = 0.9998\ ^{+0.0034}_{-0.0012} \qquad \text{and} \qquad M_H < 1002\ \text{GeV},$$

which is indeed very close to unity. This is a confirmation of the SU(2) structure of the theory and we will continue giving ρ the value unity.

12.3 Neutral-current cross sections

It is perhaps more transparent to discuss the various cross sections in the parton model (Sehgal, 1973; Kim *et al.*, 1981). The effective Lagrangian density was written, at the beginning of this chaper, in terms of the first generation of quarks. We re-express the effective interaction in terms of chiral couplings,

$$\begin{aligned}\mathcal{L} = -\frac{G}{\sqrt{2}}\bar{\nu}\gamma^\mu(1-\gamma_5)\nu\big\{&\bar{u}\gamma_\mu[u_L(1-\gamma_5) + u_R(1+\gamma_5)]u \\ &+ \bar{d}\gamma^\mu[d_L(1-\gamma_5) + d_R(1+\gamma_5)]d + \cdots\big\},\end{aligned} \tag{12.20}$$

with u_L and u_R the couplings of the left- and right-handed up quarks and with a similar definition for d_L and d_R. The ellipses indicate again contributions from higher generations. We adopted this notation because it is convenient to write down the elementary cross sections as they were classified in Section 8.3 in terms of the chiralities of the leptonic and hadronic vertices. The new couplings are related to those defined at the beginning of this chapter as follows:

$$\begin{aligned} u_{\rm L} &= \frac{1}{4}(x+y+\gamma+\delta), \\ u_{\rm R} &= \frac{1}{4}(x-y+\gamma-\delta), \\ d_{\rm L} &= \frac{1}{4}(-x-y+\gamma+\delta), \\ d_{\rm R} &= \frac{1}{4}(-x+y+\gamma-\delta). \end{aligned} \tag{12.21}$$

The neutrino experiments determined combinations of $u_{\rm L}, \ldots, d_{\rm R}$, which then were translated into $x,\ y,\ \gamma$, and δ, thus testing the isospin and parity content of the current. Finally, they were all determined in terms of a single mixing angle $\sin^2\theta_{\rm W}$. The expressions become rather long and it is convenient to introduce a shorter notation. We denote generically by f_q and $f_{\bar q}$ the parton distribution functions for q and $\bar q$ and their left-handed or right-handed couplings by $q_{\rm L}$ and $q_{\rm R}$, respectively. One easily finds cross sections for the elementary processes

$$\begin{aligned} \frac{{\rm d}\sigma_{\rm NC}(\nu {\rm q})}{{\rm d}x\,{\rm d}y} &= \frac{2G^2ME}{\pi}xf_q(x)\left[q_{\rm L}^2+q_{\rm R}^2(1-y)^2\right], \\ \frac{{\rm d}\sigma_{\rm NC}(\bar\nu {\rm q})}{{\rm d}x\,{\rm d}y} &= \frac{2G^2ME}{\pi}xf_q(x)\left[q_{\rm R}^2+q_{\rm L}^2(1-y)^2\right], \\ \frac{{\rm d}\sigma_{\rm NC}(\nu \bar{\rm q})}{{\rm d}x\,{\rm d}y} &= \frac{2G^2ME}{\pi}xf_{\bar q}\left[q_{\rm R}^2+q_{\rm L}^2(1-y)^2\right], \\ \frac{{\rm d}\sigma_{\rm NC}(\bar\nu \bar{\rm q})}{{\rm d}x{\rm d}y} &= \frac{2G^2ME}{\pi}xf_{\bar q}\left[q_{\rm L}^2+q_{\rm R}^2(1-y)^2\right]. \end{aligned} \tag{12.22}$$

There are various ways to combine these cross sections and isolate the coupling constants. In experiments with isoscalar targets,

$$f_{\rm u}(x) = f_{\rm d}(x) = f(x) \qquad \text{and} \qquad f_{\bar{\rm u}}(x) = f_{\bar{\rm d}}(x) \equiv \bar f(x).$$

Furthermore, we can integrate over y and set $K = 2G^2ME/\pi$ to obtain

$$\begin{aligned} \frac{{\rm d}\sigma_{\rm NC}(\nu {\rm N})}{{\rm d}x} &= Kx\left[\left(f+\frac{1}{3}\bar f\right)\left(u_{\rm L}^2+d_{\rm L}^2\right)+\left(\frac{1}{3}f+\bar f\right)\left(u_{\rm R}^2+u_{\rm R}^2\right)\right], \\ \frac{{\rm d}\sigma_{\rm NC}(\bar\nu {\rm N})}{{\rm d}x} &= Kx\left[\left(\frac{1}{3}f+\bar f\right)\left(u_{\rm L}^2+d_{\rm L}^2\right)+\left(f+\frac{1}{3}\bar f\right)\left(u_{\rm R}^2+d_{\rm R}^2\right)\right], \\ \frac{{\rm d}\sigma_{\rm CC}(\nu {\rm N})}{{\rm d}x} &= Kx\left(f(x)+\frac{1}{3}\bar f\right), \\ \frac{{\rm d}\sigma_{\rm CC}(\bar\nu {\rm N})}{{\rm d}x} &= Kx\left(\frac{1}{3}f+\bar f\right). \end{aligned} \tag{12.23}$$

Most experiments measure ratios of cross sections, where the flux of the neutrinos drops out. A popular ratio is

$$R_\nu = \frac{\sigma_{NC}(\nu N)}{\sigma_{CC}(\nu N)} = \left(u_L^2 + d_L^2\right) + \frac{2-B}{2+B}\left(u_R^2 + d_R^2\right), \tag{12.24}$$

with

$$B = \frac{\int_0^1 dx\, x\left[f(x) - \bar{f}(x)\right]}{\int_0^1 dx\, x\left[f(x) + \bar{f}(x)\right]} \tag{12.25}$$

measuring the relative strength of the valence- and the sea-quark contributions. For instance, $B = 1$ corresponds to vanishing sea contribution. For the experimental value $B = 0.8$ the ratio becomes

$$R_\nu = \frac{1}{2} - \sin^2\theta_W + \frac{50}{63}\sin^4\theta_W. \tag{12.26}$$

The experimental values for R_ν and $R_{\bar{\nu}}$ are

$$R_\nu = 0.29 \pm 0.01 \qquad \text{and} \qquad R_{\bar{\nu}} = 0.34 \pm 0.03.$$

In order to compare them with the prediction of Eq. (12.23) it is necessary to include precise quark distribution functions. They include contributions from sea-quark (s, $\bar{\text{s}}$) and (c, $\bar{\text{c}}$) pairs of the target. In addition, scaling violations, which have been established and analyzed in charged-current reactions, must be included (Kim *et al.*, 1981). The analysis yields

$$u_L^2 + d_L^2 = 0.29 \pm 0.01 \qquad \text{and} \qquad u_R^2 + d_R^2 = 0.03 \pm 0.01,$$

which lead to the value $\sin^2\theta_W = 0.228 \pm 0.001$.

12.4 Parity violation in electron scattering

Effects of weak neutral currents in low-energy ($Q^2 \ll M_Z^2$) electron–hadron reactions are submerged in the dominant electromagnetic interaction. For these reactions we must search for a clear signature of weak origin, such as parity violation. Experiments of this type have been carried out in deep inelastic electron–hadron scattering and in atomic physics (see Problem 4). The couplings of the Z boson to electrons and quarks have been discussed already.

A parity-violating observable is the difference of cross sections for right- and left-handed polarized electrons. These are electrons polarized along their direction of motion, i.e. electrons with definite helicity. Since helicity changes sign under spatial reflection, a difference between the two cross sections is an indication of parity violation. We denote the left-handed and right-handed electrons by the spinors

$$e_{\mathrm{L,R}} = \frac{1}{2}(1 \mp \gamma_5)u(k), \tag{12.27}$$

respectively. Their interactions at high energies with protons and neutrons are described with sufficient accuracy by the parton model. Consequently, we can write the hadronic neutral current as

$$J_\mu(x) = \bar{u}\gamma_\mu[u_{\mathrm{L}}(1-\gamma_5) + u_{\mathrm{R}}(1+\gamma_5)]u + \bar{d}\gamma_\mu[d_{\mathrm{L}}(1-\gamma_5) + d_{\mathrm{R}}(1+\gamma_5)]d \tag{12.28}$$

as it appears in Eq. (12.20). The interaction of the electrons with hadrons now involves the exchange of a photon and a Z boson. In cross sections there are contributions from the electromagnetic amplitude and weak terms. The latter contribution is responsible for the asymmetry. The amplitudes are given as

$$m_\gamma = -\frac{\mathrm{i}e^2}{q^2}\bar{e}\gamma_\mu e\,(e_{\mathrm{u}}\bar{u}\gamma^\mu u + \cdots), \tag{12.29}$$

and

$$\begin{aligned} m_{\mathrm{Z}} = -&\frac{\mathrm{i}g^2}{\cos^2\theta_{\mathrm{W}}\left(q^2 - M_{\mathrm{Z}}^2\right)}\left(g_{\mathrm{L}}\bar{e}_{\mathrm{L}}\gamma_\mu e_{\mathrm{L}} + g_{\mathrm{R}}\,\bar{e}_{\mathrm{R}}\gamma_\mu e_{\mathrm{R}}\right) \\ &\times\left[u_{\mathrm{L}}\bar{u}\gamma^\mu(1-\gamma_5)u + u_{\mathrm{R}}\bar{u}\gamma^\mu(1+\gamma_5)u + \cdots\right], \end{aligned} \tag{12.30}$$

where e_{u} is the charge of the up quark and the ellipses indicate contributions from other quarks. For the Z-boson couplings, g_{L} and g_{R} are the helicity couplings to electrons while u_{L} and u_{R} are the corresponding couplings to the up quark. For the computation of the interference term we follow the presentation of Section 8.3, where it was shown that, in the squared amplitude, the following conditions hold.

(i) The electron bilinears are left-handed or right-handed. The same is true for the quarks.
(ii) When left-handed leptonic couplings combine with left-handed quark couplings then $\mathrm{d}\sigma/\mathrm{d}y$ is independent of y. The same holds for right-handed leptonic couplings with right-handed hadronic combinations.
(iii) When left-handed leptonic couplings combine with right-handed hadronic couplings, then the dependence is $(1-y)^2$.

The form of the interference terms now follows:

$$\frac{\mathrm{d}\sigma_{\mathrm{L}}}{\mathrm{d}x\,\mathrm{d}y} \propto \left[g_{\mathrm{L}}u_{\mathrm{L}} + g_{\mathrm{L}}u_{\mathrm{R}}(1-y)^2\right]u(x) + \cdots, \tag{12.31}$$

$$\frac{\mathrm{d}\sigma_{\mathrm{R}}}{\mathrm{d}x\,\mathrm{d}y} \propto \left[g_{\mathrm{R}}u_{\mathrm{L}}(1-y)^2 + g_{\mathrm{R}}\,u_{\mathrm{R}}\right]u(x) + \cdots, \tag{12.32}$$

with the subscripts L and R in the cross section denoting left- and right-handed polarized electrons and the ellipses indicating contributions from down quarks.

The parity-violating observable is built into the asymmetry

$$A = \frac{d\sigma_R - d\sigma_L}{d\sigma_R + d\sigma_L}, \tag{12.33}$$

which is easy to derive from (12.31) and (12.32). To arrive at the final result, we must include the down quarks. In an isoscalar target, such as deuterium or carbon, there are equal numbers of up and down quarks, so only the combination $u(x) + d(x)$ appears in the cross sections, which drops out in the asymmetry. On collecting the various terms together, the asymmetry is expected to be

$$A = \frac{GQ^2}{\sqrt{2}4\pi\alpha}\frac{9}{5}\left[a_1 + a_2\frac{1-(1-y)^2}{1+(1+y)^2}\right], \tag{12.34}$$

with $a_1 = 1 - (20/9)\sin^2\theta_W$ and $a_2 = 1 - 4\sin^2\theta_W$. I have given several steps of the derivation so that the interested reader can reproduce it using the various coupling constants given in the book. The magnitude of the asymmetry for $Q^2 = 1\ \text{GeV}^2$ is

$$A \approx -1.6 \times 10^{-4}.$$

The effect was observed at the Stanford Linear Accelerator Center (SLAC) (Prescott *et al.*, 1978, 1979). Electron–proton and positron–proton collisions have been extended at HERA to very large values of $Q^2 = 400$–$40\,000\ \text{GeV}^2$, at which the effects of the Z propagator are also observable.

Problems for Chapter 12

1. Make judicious use of the Schwarz inequality to prove Eqs. (12.11) and (12.14).
2. Select the Feynman rules for the electron–hadron reaction and show that the effective interaction has the form

$$H^{ep}_{eff} = \frac{G}{\sqrt{2}}\bar{e}\gamma_\mu(g_V - g_A\gamma_5)e\left[\bar{u}\gamma_1^\mu(V_u + a_u\gamma_5)u + \bar{d}\gamma^\mu(V_d + a_d)d + \cdots\right],$$

where

$$\begin{aligned} g_V &= \frac{1}{2} - 2\sin^2\theta_W, & g_A &= -\frac{1}{2}; \\ V_u &= \left(1 - \frac{8}{3}\sin^2\theta_W\right), & a_u &= -1; \\ V_d &= -\left(1 - \frac{4}{3}\sin^2\theta_W\right), & a_d &= 1; \end{aligned} \tag{12.35}$$

and the ellipses stand for strange and heavier quarks.

3. Combine the results of the previous problem with the outline of Section 12.4 and obtain the final form of the asymmetry.

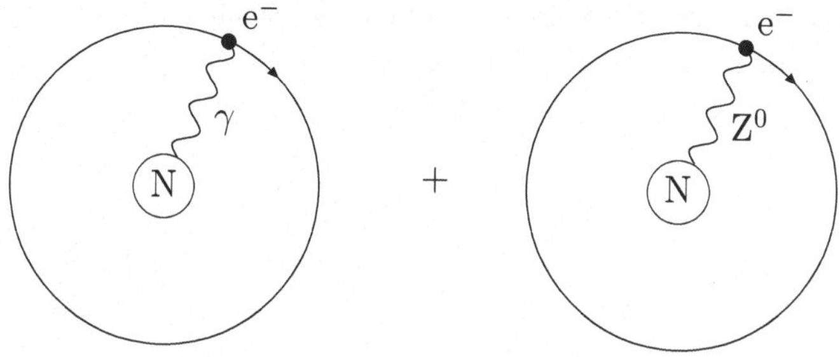

Figure 12.1. A schematic drawing of a Z exchange in an atom.

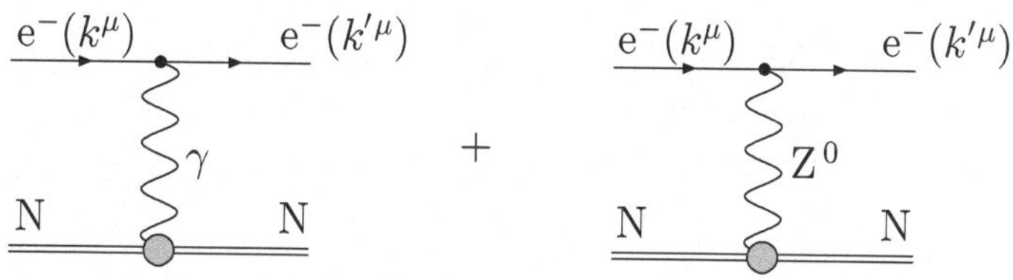

Figure 12.2. Feynman diagrams for electron–nucleus interaction.

4. Another manifestation of neutral-current interactions appears as parity violation in atoms. The neutral current introduces a new interaction between the orbiting electron and the nucleus. The total force in the atom is the sum of electromagnetic and weak diagrams (Fig. 12.1) or can be expressed in terms of Feynman diagrams (Fig. 12.2).

   The sum of the amplitudes contributes

$$m = \bar{e}(k')\gamma^{\mu}e(k)\frac{e^2}{q^2}\langle N|J_{\mu}^{\mathrm{em}}|N\rangle + \frac{g^2}{8\cos^2\theta_{\mathrm{W}}}\bar{e}(k')\gamma^{\mu}(g_{\mathrm{V}} + g_{\mathrm{A}}\gamma_5)e(k)\frac{1}{q^2 - M_{\mathrm{Z}}^2}\langle N|J_{\mu}^{\mathrm{NC}}|N\rangle,$$

where $q_{\mu} = k_{\mu} - k'_{\mu} = p'_{\mu} - p_{\mu}$. In order to identify weak effects, a signal with parity violation is required.

   Two parity-violating amplitudes are

$$M_1 = \frac{G}{\sqrt{2}}\bar{e}(k')g_{\mathrm{A}}\gamma^{\mu}\gamma_5 e(k)\left\{\langle N|V_{\mu}^3|N\rangle - 2\sin^2\theta_{\mathrm{W}}\langle N|J_{\mu}^{\mathrm{em}}|N\rangle\right\}$$

and

$$M_2 = \frac{G}{\sqrt{2}}\bar{e}(k')g_{\mathrm{V}}\gamma^{\mu}e(k)\langle N|A_{\mu}|N\rangle.$$

For $\sin^2\theta_{\mathrm{W}} = 0.25$ (which is close to the experimental value), only the M_1 amplitude survives. For this reason and because of the suppression of the hadronic matrix element in M_2, we discuss below only M_1

The momenta involved in atomic experiments are small, so it is convenient to obtain a non-relativistic limit of the weak interaction.

(i) For $|\vec{k}| \ll m_e$ and $|\vec{k}'| \ll m_e$, show that

$$\frac{1}{q^2 - M_Z^2} \approx \frac{1}{4\pi} \int d^3r\, e^{i\vec{q}\cdot\vec{r}} \frac{e^{-M_Z r}}{r}.$$

The weak interaction is of short range and, in the limit of large M_Z, it acts at the origin, where the nucleus is located. Later on we replace the Yukawa potential by a three-dimensional δ-function.

(ii) For reduction of the hadronic matrix element to the non-relativistic limit, consider

$$\langle N, p'|J_\mu^{em}|N, p\rangle = \bar{u}(p')\left[\gamma_\mu F_1(q^2) + i\sigma_{\mu\nu}\frac{(p-p')^\nu}{2M}F_2(q^2)\right]u(p).$$

Show that, in the non-relativistic limit, only the $\mu = 0$ component survives and gives the charge Q of the nucleus. Similar arguments for the V_μ^3 matrix element give

$$\langle N, p'|V_\mu^3|N, p\rangle = g_{\mu 0}\frac{1}{2}(Z - N),$$

with Z the number of protons and N on the right-hand side the number of neutrons in the nucleus. On combining the results from steps (i) and (ii), we find that the nucleus generates, through the M_1 amplitude, the potential

$$Z_\mu(r) = g_{\mu 0}\frac{G}{\sqrt{2}}\delta^3(\vec{r})Q_W(Z, N),$$

with $Q_W(Z, N) = \frac{1}{2}[Z(1 - 4\sin^2\theta_W) + N]$.

(iii) The transition-matrix element of the electron is

$$\langle e_f|H_{PV}|e_i\rangle = \int \bar{e}_f(r)\gamma_\mu\gamma_5 Z_\mu e_i(r) d^3r.$$

For atomic physics the electron wave function can be written as

$$e_i(r) = \begin{pmatrix} 1 \\ \dfrac{\vec{\sigma}\cdot\vec{\nabla}}{2m_e} \end{pmatrix} \psi_i(r),$$

with $\psi_i(r)$ the space wave function of level i, m_e the mass of the electron and $\vec{\nabla}$ the momentum operator. Show that the matrix element reduces to

$$\langle e_f|H_{PV}|e_i\rangle = \frac{G}{\sqrt{2}}\frac{1}{2m_e}Q_W \int d^3r\, \psi_f^*(r)\left[\vec{\sigma}\cdot\vec{\nabla}\delta^3(r) + \delta^3(r)\sigma\cdot\vec{\nabla}\right]\psi_i(r).$$

The effect of the neutral current on an atomic level is to induce the mixing of levels with opposite parities. Thus the absorption rates of beams of monochromatic light with various polarizations by atoms differ. One effect of parity violation in heavy atoms is the rotation of the plane of polarization of laser light passing through atomic vapors. Such experiments have been performed and effects of the neutral current have been observed (Bouchiat and Bouchiat, 1974).

References

Bouchiat, M. A., and Bouchiat, C. C. (1974), *Phys. Lett.* **B48**, 111
Kim, J. E., Langacker, P., Levine, M., and Williams, H. H. (1981), *Rev. Mod. Phys.* **53**, 211
Pais, A., and Treiman, S. B. (1972), *Phys. Rev.* **D6**, 2700
Paschos, E. A., and Wolfenstein, L. (1973), *Phys. Rev.* **D7**, 91
Prescott, C. Y., Atwood, W. B., Cottrell, R. L. A. *et al.* (1978), *Phys. Lett.* **B77**, 347 (1979) *Phys. Lett.* **B84**, 524
Sehgal, L. M. (1973), *Nucl. Phys.* **B65**, 141

13
Physics of neutrinos

13.1 Neutrino masses

Neutrinos as elementary particles have remarkable properties. They have only weak and gravitational interactions, which allows them to travel through matter making very few interactions. They carry a global quantum number, known as lepton number, which can be broken without disturbing the conservation of electric charge. The breaking of lepton number resides on the mass matrices which we introduce in this section.

Their unique properties have led to several discoveries, the newest among them being neutrino oscillations, which provide information on their mass differences and mixing parameters. Neutrino oscillations also create new questions concerning their properties, which are now under active investigation.

Masses for quarks and leptons were introduced in Chapters 8 and 9 through Yukawa couplings to the Higgs doublet. The neutrino remained massless because interactions of right-handed neutrinos had not been observed. This is a unique and unfamiliar situation, because all other fermions have right-handed components. In this chapter we shall describe the properties of neutrinos and introduce right-handed neutrinos $\mathrm{N_R}$, which are singlets under $\mathrm{SU(2)_L}$. The representation content for the electron family is

$$\Psi_\mathrm{L} = \begin{pmatrix} \nu \\ \mathrm{e} \end{pmatrix}_\mathrm{L}, \ \mathrm{N_R} \text{ and } \mathrm{e_R}, \tag{13.1}$$

with $e_\mathrm{R,L} = \frac{1}{2}(1 \pm \gamma_5)e$. We should have written the right-handed state as ν_R; however, right-handed neutrinos will play later a special role (Majorana), so we decided to denote them $\mathrm{N_R}$. There is an analogous classification of the leptonic states for the muon and tau families.

A mass term of the form

$$m_\mathrm{D}\bar{\nu}_\mathrm{L} N_\mathrm{R} + \text{h.c.} \tag{13.2}$$

is possible and is generated from the Yukawa coupling

$$\mathcal{L}_{\mathrm{Y}}^{\nu} = \sum f_{ll'} \bar{\Psi}_{\mathrm{L}}^{l} \hat{\Phi} N_{\mathrm{R}}^{l'} + \text{h.c.}, \tag{13.3}$$

where Φ is the standard Higgs doublet and

$$\hat{\Phi} = \mathrm{i}\tau_2 \Phi^* = \begin{pmatrix} \Phi_0 \\ -\Phi^- \end{pmatrix} \overrightarrow{\text{breaking}} \begin{pmatrix} \Phi_0 + v \\ 0 \end{pmatrix}. \tag{13.4}$$

We call this the Dirac mass term. The discussion so far is similar to that of quarks and charged leptons. However, the observed extreme smallness of neutrino masses seems to require a special treatment. It is possible to introduce another term, known as the Majorana mass term, which brings special properties and is a candidate for explaining the small masses.

Before we proceed with new properties of the neutrinos, it is instructive to point out which spinors correspond to the above states. The identification is helpful when we proceed with calculations. The Dirac equation has four solutions: two solutions with $E > 0$ and two with $E < 0$, which describe particle and antiparticle states to be denoted by u and v, respectively. We follow standard textbook notation with

$$\psi_{1,2} = u_i \mathrm{e}^{-\mathrm{i}p\cdot x} \qquad \text{and} \qquad \psi_{3,4} = v_i \mathrm{e}^{\mathrm{i}p\cdot x} \qquad \text{for } i = 1, 2, \tag{13.5}$$

where, for particles moving in the z-direction, the four spinors are

$$\begin{aligned} u_1 &= \sqrt{\frac{E+m}{2m}} \begin{pmatrix} 1 \\ 0 \\ \dfrac{p_z}{E+m} \\ 0 \end{pmatrix}, & u_2 &= \sqrt{\frac{E+m}{2m}} \begin{pmatrix} 0 \\ 1 \\ 0 \\ \dfrac{-p_z}{E+m} \end{pmatrix}, \\ v_1 &= \sqrt{\frac{E+m}{2m}} \begin{pmatrix} \dfrac{p_z}{E+m} \\ 0 \\ 1 \\ 0 \end{pmatrix}, & v_2 &= \sqrt{\frac{E+m}{2m}} \begin{pmatrix} 0 \\ \dfrac{-p_z}{E+m} \\ 0 \\ 1 \end{pmatrix}. \end{aligned} \tag{13.6}$$

The normalization of the spinors is as follows: $\overline{u_i} u_j = \delta_{ij}$ and $\overline{v_i} v_j = -\delta_{ij}$.

For massless neutrinos the spinors are eigenfunctions of the operator γ_5 and we define the spinor for the neutrino as

$$\nu = \left(\frac{1-\gamma_5}{2}\right) u_2 \tag{13.7}$$

and the spinor for the antineutrino as

$$\bar{\nu} = \left(\frac{1+\gamma_5}{2}\right) v_1. \tag{13.8}$$

In this limit we speak of left-handed neutrinos and right-handed antineutrinos. We also notice that in the limit $p_z \gg m$ there are only two independent spinors.

For massive particles the situation is different. Considering the Hamiltonian of a free Dirac particle,

$$H = c\vec{\alpha} \cdot \vec{p} + \beta mc^2, \tag{13.9}$$

where $\beta\vec{\alpha} = \gamma_i$ and $\beta = \gamma_0$, we notice that γ_5 does not commute with the Hamiltonian since $[\gamma_5, \gamma_0] \neq 0$. Thus, for massive neutrinos, γ_5 is not a good quantum number; i.e. the spinors are not eigenstates of the γ_5 operator. For the massive case we introduce another operator, the helicity

$$h = \frac{1}{2}\frac{\vec{p} \cdot \vec{s}}{|\vec{p}|} = \frac{1}{2}\hat{p}^i \begin{pmatrix} \sigma^i & 0 \\ 0 & \sigma^i \end{pmatrix}. \tag{13.10}$$

It is easy to verify that the spinors u_1 und v_1 are eigenfunctions with helicity $\frac{1}{2}$. Similarly, u_2 und v_2 are eigenfunctions of the helicity operator with eigenvalues $-\frac{1}{2}$. In a given Lorentz frame, helicity in a reaction is conserved. However, the helicity of a massive particle depends on the frame, because, by moving very fast, we can reverse the momentum of a particle, leaving its spin unchanged.

A new mass term of the form

$$\frac{1}{2}M_{\mathrm{m}}\bar{N}_{\mathrm{R}}^{\mathrm{c}}N_{\mathrm{R}} + \mathrm{h.c.} \tag{13.11}$$

is allowed to be present, since it is Lorentz- and SU(2)$_{\mathrm{L}}$-invariant. This is known as a Majorana mass term and is unique to neutrinos, which are neutral particles. A Majorana mass term conserves electric charge but changes the lepton number by two units. It may be introduced to the Langrangian as an additional term or as a new interaction term, coupled to a new scalar particle which is an SU(2)$_{\mathrm{L}}$ singlet. A Majorana mass is generated by assigning to the new scalar particle a vacuum expectation value. Consequently, we generate Dirac mass terms through Eq. (13.3) and a Majorana term through Eq. (13.11). On collecting these terms together, one obtains the neutrino mass matrix:

$$\begin{pmatrix} \bar{\nu}_{\mathrm{L}} & \bar{N}_{\mathrm{R}}^{\mathrm{c}} \end{pmatrix} \begin{pmatrix} 0 & m_{\mathrm{D}} \\ m_{\mathrm{D}}^{\mathrm{T}} & M_{\mathrm{m}} \end{pmatrix} \begin{pmatrix} \nu_{\mathrm{L}} \\ N_{\mathrm{R}} \end{pmatrix}. \tag{13.12}$$

Once we have introduced mass terms that do not preserve the original symmetry, we must solve the problem again using the Lagrangian. This means that, after introducing Dirac and/or Majorana mass terms, we should solve the problem using the rules of the new Lagrangian (see Problem 13.3). For our specific case, we can no longer use the eigenfunctions of Eq. (13.6). We must diagonalize the mass matrix and use the new mass eigenstates (physical states) which will have components that couple the various flavors with each other and, in addition, couple particles to antiparticles (when Majorana mass terms are present).

For $M_{\rm m} \gg m_{\rm D}$ the mass matrix of the form given in Eq. (13.12) has eigenvalues with specific properties: one eigenvalue is large and the other is suppressed. This method of introducing masses for neutrinos is known as the see-saw mechanism, to which we shall return in the fourth section. In the meantime we shall discuss neutrino oscillations in free space and in matter, which are active fields of research nowadays.

13.2 Neutrino oscillations

The couplings of Dirac neutrinos to charged and neutral currents conserve the lepton number, as has been tested in many experiments. Lepton flavor can be violated in the mass matrix, with the appearance of off-diagonal elements. In this case the eigenfunctions of the Hamiltonian are superpositions of neutrinos with various flavor numbers. We shall call them the mass eigenstates and they have the time development given below in Eq. (13.14). In a physical reaction, however, the neutrinos that are produced have definite flavor number. Their time development requires special attention because we must rewrite the flavor states in terms of mass eigenstates, whose time development is that given in Eq. (13.14). This mis-match between the production of flavor states and the time development of mass states leads to an oscillation of lepton quantum numbers that will be described below.

We demonstrate the mixing phenomenon for two generations of neutrinos, for which the algebra is simpler. We consider $\nu_{\rm e}$ and ν_{μ} neutrinos with the mass matrix

$$i\frac{\partial}{\partial t}\begin{pmatrix}\nu_{\rm e}\\ \nu_{\mu}\end{pmatrix} = \mathcal{H}_{\rm mass}\begin{pmatrix}\nu_{\rm e}\\ \nu_{\mu}\end{pmatrix}, \quad \text{with} \quad \mathcal{H}_{\rm mass} = \begin{pmatrix} m_{\rm ee} & m_{\rm e\mu}\\ m_{\mu\rm e} & m_{\mu\mu}\end{pmatrix}. \tag{13.13}$$

The mass matrix mixes electron and muon neutrinos in a manner analogous to atomic physics, where energy levels are mixed through the interactions with external fields, such as magnetic fields. For the sake of simplicity, we assume that the matrix elements are real and in addition that $m_{\mu\rm e} = m_{\rm e\mu}$. A symmetric matrix is diagonalized by an orthogonal matrix to be denoted by O and has the eigenvalues m_1 and m_2. A mass eigenstate with momentum p has the time development

$$|\nu_i(t)\rangle = |\nu_i(0)\rangle e^{-iE_i t}, \tag{13.14}$$

with $i = 1$ or 2 and $E_i = \sqrt{p^2 + m_i^2}$. To avoid confusion, we shall use two types of subscripts, with numbers denoting physical eigenstates and letters denoting flavor states. Let us parametrize the orthogonal matrix as follows:

$$O = \begin{pmatrix} \cos\theta & \sin\theta \\ -\sin\theta & \cos\theta \end{pmatrix};$$

then the mass eigenstates are given by

$$\begin{pmatrix} \nu_1(t) \\ \nu_2(t) \end{pmatrix} = \begin{pmatrix} \cos\theta & -\sin\theta \\ \sin\theta & \cos\theta \end{pmatrix} \begin{pmatrix} \nu_e(t) \\ \nu_\mu(t) \end{pmatrix}. \tag{13.15}$$

We can easily invert this equation to obtain

$$\begin{pmatrix} \nu_e(t) \\ \nu_\mu(t) \end{pmatrix} = \begin{pmatrix} \cos\theta & \sin\theta \\ -\sin\theta & \cos\theta \end{pmatrix} \begin{pmatrix} \nu_1(t) \\ \nu_2(t) \end{pmatrix}, \tag{13.16}$$

which, with the help of Eq. (13.14), can be rewritten as

$$\begin{aligned} |\nu_e(t)\rangle &= e^{-iE_1t}|\nu_1(0)\rangle\cos\theta + e^{-iE_2t}|\nu_2(0)\rangle\sin\theta, \\ |\nu_\mu(t)\rangle &= -e^{-iE_1t}|\nu_1(0)\rangle\sin\theta + e^{-iE_2t}|\nu_2(0)\rangle\cos\theta. \end{aligned} \tag{13.17}$$

The general structure of these equations is summarized by

$$|\nu_\alpha(t)\rangle = \sum_{k=1,2} U_{\alpha k} e^{-iE_kt}|\nu_k(0)\rangle,$$

with the unitary matrix $U_{\alpha k}$ defining the mixing between flavor and mass states. The mixing matrix was introduced a long time ago and is referred to as the MNS matrix (Maki *et al.*, 1962).

On replacing $\nu_1(0)$ and $\nu_2(0)$ by the flavor eigenstates given in Eq. (13.15), we obtain the final result

$$\begin{aligned} |\nu_e(t)\rangle &= (\cos^2\theta\, e^{-iE_1t} + \sin^2\theta\, e^{-iE_2t})|\nu_e(0)\rangle \\ &\quad + \sin\theta\cos\theta(e^{-iE_2t} - e^{-iE_1t})|\nu_\mu(0)\rangle, \\ |\nu_\mu(t)\rangle &= \sin\theta\cos\theta(e^{-iE_2t} - e^{-iE_1t})|\nu_e(0)\rangle \\ &\quad + (\cos^2\theta\, e^{-iE_1t} + \sin^2\theta\, e^{-iE_2t})|\nu_\mu(0)\rangle. \end{aligned} \tag{13.18}$$

This shows explicitly that the flavor content of the wave function changes with time. For example, at $t = 0$ the first equation contains only an electron neutrino. In the course of time a ν_μ component develops. The second equation describes a state which starts as $\nu_\mu(0)$.

We consider a state that starts as $\nu_e(0)$ and compute the probability of finding a $\nu_e(t)$. The masses of the neutrinos are small relative to their momenta and one can use the approximation $E_i = \sqrt{p^2 + m_i^2} \approx E[1 + m_i^2/(2E^2)]$ to arrive at

$$P_{ee}(t) = |\langle \nu_e | \nu_e(t) \rangle|^2 = 1 - \sin^2(2\theta)\sin^2\left(\frac{\Delta m^2}{4E}t\right), \tag{13.19}$$

with $\Delta m^2 = m_2^2 - m_1^2$. Similarly, we compute the probability of finding a ν_μ at time t,

$$P_{\mu e}(t) = |\langle \nu_\mu | \nu_e(t) \rangle|^2 = \sin^2(2\theta)\sin^2\left(\frac{\Delta m^2}{4E}t\right). \tag{13.20}$$

The sum of the two probabilities is equal to unity. A mono-energetic neutrino beam thus oscillates with amplitude $\sin^2(2\theta)$ and wave-number $\Delta m^2/(4E)$. For oscillations to occur, we need a non-zero θ and at least one non-zero mass. The amplitude is maximal for $\theta = \pi/4$.

In vacuum-oscillation experiments there is a redundancy in the values of the mixing angle. The same numerical value of $\sin^2(2\theta)$ appears eight times when θ varies between zero and 2π. This means that in oscillation experiments it suffices to consider angles in the interval between zero and $\pi/4$. In other situations for which the mixing depends on $\cos(2\theta)$ the range for the mixing angle must be extended from 0 to $\pi/2$, as we shall discuss at the end of the next section.

We can express the various quantities in useful units and substitute $t \simeq L$, where the speed of light c is set equal to unity,

$$\frac{\Delta m^2 L}{4E} = 1.27 \frac{L}{\text{km}} \frac{\Delta m^2}{\text{eV}^2} \frac{\text{GeV}}{E}. \tag{13.21}$$

This entity is a dimensionless quantity and we are free to use the units. The units in Eq. (13.21) are now standard and convenient for terrestrial experiments. This formula defines the sensitivity of an experiment, since oscillations occur when $\Delta m^2 L/E$ is of order unity. As a function of the baseline length, the maximum of the oscillation occurs at $L \simeq 2E/\Delta m^2$.

The generalization to three or more families is straightforward and leads to the transition probability

$$|\langle \nu_\beta | \nu_\alpha(t) \rangle|^2 = \delta_{\alpha\beta} - 2\,\text{Re}\left\{ \sum_{j>i} U_{\alpha i} U_{\alpha j}^* U_{\beta i}^* U_{\beta j} \left[1 - \exp\left(-\text{i}\frac{\Delta m_{ij}^2}{2E} L \right) \right] \right\}. \tag{13.22}$$

For two neutrino families, $U_{\alpha i}$ is a simple 2×2 orthogonal matrix. Upon substitution the probabilities simplify to Eqs. (13.19) and (13.20). For three families a hierarchy of the form $\Delta m_{12}^2 \gg \Delta m_{13}^2 \simeq \Delta m_{23}^2$ and the smallness of one mixing-matrix element (U_{e3}) describes all existing experiments.

Finally, we present a simple description of what happens in oscillations. A $|\nu_\alpha\rangle$ is composed of various waves with different m_i. At a given energy the heavier mass states oscillate faster and the various $|\nu_i\rangle$ components come out of phase, so at a certain distance they do not sum up to a $|\nu_\alpha\rangle$. Since the oscillatory term comes from an interference between the different m_i, a common phase factor of the $|\nu_i\rangle$s plays no role and can be ignored. This will be useful in the next section, when we turn to oscillation in matter.

13.2.1 Oscillation in matter

The neutral-current interaction of neutrinos with matter is extremely weak, but can nonetheless affect oscillations (Wolfenstein, 1978). The reason is that the momentum transfer to the target can have a large wavelength, such that the neutrino interacts coherently with all the particles within its wavelength. At the same time the difference $E_1 - E_2 \simeq \Delta m_{12}^2/(2E)$ may correspond to a wavelength of the same magnitude, so the time development of the mass eigenstates is influenced by the interactions with the medium. In the medium the flavor eigenstates ν_e and ν_μ interact with electrons and protons with different cross sections and modify the development of the mass eigenstates.

The Hamiltonian on the mass basis is given by

$$\mathrm{i}\frac{\partial \Psi_i}{\partial t} = \mathcal{H}_0 \Psi_i = E_i \Psi_i \simeq p\left(1 + \frac{m_i^2}{2p^2}\right)\Psi_i \to \frac{m_i^2}{2E}\Psi_i, \tag{13.23}$$

where after the arrow we omitted the term proportional to the unit matrix. This term does not influence the oscillations because it can be eliminated by the transformation $\nu_i \to \nu_i \mathrm{e}^{-\mathrm{i}pt}$ common for all neutrinos. For this reason we can simplify the Hamiltonian,

$$\mathcal{H}_0 \Psi = \mathcal{H}_0 \begin{pmatrix} \nu_1 \\ \nu_2 \end{pmatrix} = \frac{1}{2E}\begin{pmatrix} m_1^2 & 0 \\ 0 & m_2^2 \end{pmatrix}\begin{pmatrix} \nu_1 \\ \nu_2 \end{pmatrix}. \tag{13.24}$$

The interactions with matter, however, involve flavor states. For this reason we transform the equations to the flavor basis where we include the interactions with matter. The solution of the new eigenvalue problem describes the propagating eigenfunctions. In the flavor basis

$$\mathcal{H}_{\text{flavor}} \equiv \frac{1}{2E} U \begin{pmatrix} m_1^2 & 0 \\ 0 & m_2^2 \end{pmatrix} U^\dagger = \frac{1}{4E}\begin{pmatrix} \Sigma - \Delta m^2 \cos(2\theta) & \Delta m^2 \sin(2\theta) \\ \Delta m^2 \sin(2\theta) & \Sigma + \Delta m^2 \cos(2\theta) \end{pmatrix}, \tag{13.25}$$

where $\Sigma = m_1^2 + m_2^2$, $\Delta m^2 = m_1^2 - m_2^2$, and U is the mixing matrix in Eq. (13.16).

In a medium there is an additional interaction Hamiltonian created by the neutral and charged currents. Recall the interaction term from Chapter 8,

$$\mathcal{H}_{\rm int} = \frac{G_{\rm F}}{\sqrt{2}}\,\bar{\psi}_x\gamma_\mu\left(g_{\rm V}^x - g_{\rm A}^x\gamma_5\right)\psi_x\bar{\psi}_{\nu_{\rm e}}\gamma^\mu(1-\gamma_5)\psi_{\nu_{\rm e}}. \tag{13.26}$$

Here x denotes weakly interacting particles in the medium, which in the Sun are electrons, neutrons, and protons. For left-handed neutrinos $(1-\gamma_5)\psi_\nu = 2\psi_\nu$. The term $\langle\bar{\psi}_x\gamma_\mu\gamma_5\psi_x\rangle$ reduces in the non-relativistic limit to the expectation value for the spin operator. In an unpolarized medium, the states ψ_x occupy all possible spin states and this term averages to zero. For the remaining term $\langle\bar{\psi}_x\gamma_\mu\psi_x\rangle$ the space components are proportional to the momentum of the particles, which is small. Consequently only the term $\langle\bar{\psi}_x\gamma_0\psi_x\rangle = n_x$ survives and is equal to the density of the particles n_x. This term produces the potential $g_{\rm V}^x\sqrt{2}G_{\rm F}n_x\psi_{\nu_{\rm e}}^\dagger\psi_{\nu_{\rm e}}$ which is added to the Hamiltonian. Finally, we obtain the Schrödinger equation from the Euler–Lagrange equation for $\psi_{\nu_{\rm e}}^\dagger$.

We already know the couplings of neutrinos to electrons from Section 8.3, where they are summarized in Table 8.1. The couplings to nucleons are obtained in a similar way:

$$\begin{aligned}
g_{\rm V}^{\rm e} &= \tfrac{1}{2} + 2\sin^2\theta_{\rm W} && \text{for } \nu_{\rm e},\\
g_{\rm V}^{\rm e} &= -\tfrac{1}{2} + 2\sin^2\theta_{\rm W} && \text{for } \nu_\mu,\\
g_{\rm V}^{\rm p} &= \tfrac{1}{2} - 2\sin^2\theta_{\rm W} && \text{for } \nu_\alpha,\\
g_{\rm V}^{\rm n} &= -\tfrac{1}{2} && \text{for } \nu_\alpha,
\end{aligned} \tag{13.27}$$

with α being either the electron neutrino or the muon neutrino. The difference of 1 in the $g_{\rm V}^{\rm e}$ contribution between $\nu_{\rm e}$ and ν_μ comes from the Fierz-transformed charged-current term in the Hamiltonian. In an electrically neutral medium $n_{\rm e} = n_{\rm p}$, therefore

$$V = \sqrt{2}G_{\rm F} \times \begin{cases} -\tfrac{1}{2}n_{\rm n} + n_{\rm e} & \text{for } \nu_{\rm e},\\ -\tfrac{1}{2}n_{\rm n} & \text{for } \nu_\mu. \end{cases} \tag{13.28}$$

The new contribution from the scattering with matter must be added to the ee and μμ elements of $\mathcal{H}_{\rm flavor}$, yielding a propagation equation

$$\mathrm{i}\frac{\mathrm{d}}{\mathrm{d}t}\begin{pmatrix}\nu_{\rm e}\\ \nu_\mu\end{pmatrix} = \frac{1}{2E}\begin{pmatrix} m_{\rm ee}^2 + 2\sqrt{2}EG_{\rm F}n_{\rm e} & m_{{\rm e}\mu}^2\\ m_{{\rm e}\mu}^2 & m_{\mu\mu}^2\end{pmatrix}\begin{pmatrix}\nu_{\rm e}\\ \nu_\mu\end{pmatrix}, \tag{13.29}$$

with matrix elements

$$\begin{aligned} m_{ee}^2 &= -\tfrac{1}{2}\,\Delta m^2 \cos(2\theta), \\ m_{\mu\mu}^2 &= \tfrac{1}{2}\,\Delta m^2 \cos(2\theta), \\ m_{e\mu}^2 &= \tfrac{1}{2}\,\Delta m^2 \sin(2\theta). \end{aligned} \tag{13.30}$$

One sees that effects of matter become more important at higher energy. Note that we ignored terms proportional to the unit matrix. Apart from the aforementioned fact that they can be transformed away, there is another justification: after diagonalization the mixing angle and the difference between the eigenvalues are independent of equal diagonal terms (see Problem 13.1). Exactly such terms appear in the formula (13.30) for the oscillation probability and have been omitted. Diagonalization of the matrix gives the mass difference

$$m_{2M}^2 - m_{1M}^2 = \Delta m_M^2 = \Delta m^2 \sqrt{[A - \cos(2\theta)]^2 + \sin^2(2\theta)}, \tag{13.31}$$

where

$$A \equiv \frac{2\sqrt{2} E G_F n_e}{\Delta m^2},$$

and the mixing angle (Wolfenstein, 1978; Mikheyev and Smirnov, 1985)

$$\tan(2\theta_M) = \frac{\sin(2\theta)}{\cos(2\theta) - A} \Rightarrow \sin^2(2\theta_M) = \frac{\sin^2(2\theta)}{[\cos(2\theta) - A]^2 + \sin^2(2\theta)}. \tag{13.32}$$

In the limit $n_e \to 0$, we recover the formulas for oscillation in vacuum. The careful reader will notice that there is a resonance effect when $A = \cos(2\theta)$ (Mikheyev and Smirnov, 1985). For this value of A the mixing in the medium is maximal, even though the mixing in the vacuum can be very small.

The angle θ_M expresses the matter eigenstates in terms of the flavor states

$$\begin{pmatrix} \nu_{1M} \\ \nu_{2M} \end{pmatrix} = \begin{pmatrix} \cos\theta_M & -\sin\theta_M \\ \sin\theta_M & \cos\theta_M \end{pmatrix} \begin{pmatrix} \nu_e \\ \nu_\mu \end{pmatrix}. \tag{13.33}$$

We discuss two cases realized in the Sun. We know that electron-type neutrinos are created in the interior of the Sun, where n_e is very large. In that region

$$\tan(2\theta_M) \simeq -\frac{2\theta}{A} \quad \text{and} \quad \theta_M \simeq \frac{\pi}{2}, \tag{13.34}$$

since $\tan(2\theta_M)$ approaches zero from negative values. Equation (13.33) now gives $\nu_{1M} \simeq -\nu_\mu$ and $\nu_{2M} \simeq \nu_e$, i.e. in the interior of the Sun the state with the heavier effective mass is the electron neutrino. As the beam transverses the Sun, n_e decreases and $\tan(2\theta_M) \simeq \tan(2\theta)$, therefore $\theta_M \simeq \theta$ and $\nu_{2M} \simeq \nu_\mu$. The flavor content of the beam changes. This phenomenon is called the MSW effect. It will be interesting to measure in terrestrial experiments the lepton-number content of the neutrinos arriving from the Sun.

Next we discuss terrestrial experiments with a ν_μ beam going through the Earth. This situation can be realized in long-baseline accelerator experiments or via atmospheric neutrinos. We assume that the ν_μ oscillates into another flavor state, as reported from the Superkamiokande experiment. Two cases are of interest.

Case 1

The ν_μ mixes with the ν_τ. The interaction with matter proceeds with the exchange of a Z^0 boson and is identical for μ and τ neutrinos. It will thus add the same term to the diagonal elements of $\mathcal{H}_{\text{flavor}}$, which does not affect the oscillation. This means that no effects of matter appear.

Case 2

The ν_μ mixes with a sterile neutrino ν_s. The interaction with matter in the transition $\nu_\mu \to \nu_\mu$ proceeds through the exchange of a Z^0 boson, but there is no such process for the ν_s. We must add an interaction with matter only in the $\mu\mu$ element and the propagation Hamiltonian has the form

$$\mathcal{H}_{\text{flavor}} = \frac{1}{2E}\begin{pmatrix} m^2_{\mu\mu} - \sqrt{2}EG_F n_n & m^2_{\mu s} \\ m^2_{\mu s} & m^2_{ss} \end{pmatrix}. \qquad (13.35)$$

The notation is analogous to that in Eq. (13.30) with mass elements $m^2_{\mu\mu}$, $m^2_{\mu s}$, and m^2_{ss} similar to those in Eq. (13.30). The angle θ now governs the oscillation between ν_μ and ν_s. We note that the sign of the matter term is reversed and the neutron density n_n replaces the density of electrons.

Finally, Eqs. (13.31) and (13.32) depend on $\cos(2\theta)$ and the range $0 \leq \theta \leq \pi/4$ does not cover negative values of $\cos(2\theta)$. The range must now be extended to $0 \leq \theta \leq \pi/2$.

13.3 Experimental results

There has been a long search for neutrino oscillations. The experiments were carried out with neutrino beams from nuclear reactors, the Sun, and accelerators, and recently with atmospheric neutrinos. In recent years several experiments have provided evidence for neutrino oscillation. We will not discuss the experiments in detail but only summarize the main results. For an overview of the current status, see Mohapatra *et al.* (2005).

- *Atmospheric neutrinos* Compelling evidence for oscillation comes from experiments with atmospheric neutrinos. They are produced in the atmosphere by meson decays, which in turn are produced by the interaction of cosmic rays with nuclei of the atmosphere. Both electron and muon neutrinos are produced in pion and muon decays and the ratio at production is roughly two muon neutrinos to one electron neutrino. Neutrino interactions are detected in a huge underground target/detector – the Superkamiokande. The experiment established that there is a decrease in the number of muon-type neutrinos and no decrease in number of electron neutrinos. This implies that the ν_μ neutrinos oscillate into another type, such as τ or sterile (see below) neutrinos. A zenith-angle dependence of the disappearance is observed, whereby more of the neutrinos coming from underneath the detector disappear. They are produced in the atmosphere and travel through the Earth in order to reach the detector. These observations require a mass difference of $\Delta m^2 \simeq 2 \times 10^{-3}$ eV^2 and the mixing angle to be maximal.
- *Solar neutrinos* Nuclear reactions in the Sun produce a tremendous amount of heat, which is radiated and at the same time produces electron neutrinos. The cycles which produce them have been studied and the neutrino spectra have been calculated. For three decades experiments measuring the flux of neutrinos from the Sun have observed deficits, indicating that ν_e neutrinos, during their journey from Sun to Earth, oscillate into other types of neutrinos. Previous experiments trying to verify this result with reactor neutrinos failed to find a deficit, indicating that the distance of 1 km from the reactors is too small. The explanation of the solar deficit requires that it is either oscillation in vacuum with a large mixing angle and $\Delta m^2 \simeq 10^{-10}$ eV^2 or oscillations inside the Sun where effects of matter according to the MSW effect are important. The latter possibility gives two solutions with either large mixing (LAMSW) or small mixing (SAMSW) and with $\Delta m^2 \simeq 8 \times 10^{-5}$ eV^2.
- *Reactor experiments* Reactor experiments search for a decrease of the antineutrino flux far away from the reactor. The CHOOZ collaboration detected $\bar{\nu}_e$ at a distance of 1 km away but saw no deficit. These data restrict the mixing angle θ_{13}, as we will discuss below. The experiment KamLAND measured the flux of $\bar{\nu}_e$ neutrinos from distant nuclear reactors. The experiment uses a target containing 1000 tons of liquid scintillator viewed by more than 1800 light-detecting photomultiplier tubes. It detects electron antineutrinos emitted by ~70 nuclear reactors in Japan and South Korea arriving from an average distance of 180 km. The ratio of the observed antineutrino interactions to the expected number without disappearance of $\bar{\nu}_e$ is 0.611 ± 0.085 (statistical uncertainty) ± 0.041 (systematic uncertainty) for $\bar{\nu}_e$ energies >3.4 MeV. In the two-flavor analysis only the "large-mixing-angle" solution is allowed. The best fit of their data gives $\Delta m_{12}^2 = 6.9 \times 10^{-5}$ eV^2 and $\sin^2(2\theta) = 1.0$, which selects the large-mixing-angle solution.
- *The Sudbury experiment* The Sudbury Neutrino Observatory (SNO) detects ^{8}B solar neutrinos through the reactions

$$\nu_e + \mathrm{d} \rightarrow \mathrm{p} + \mathrm{p} + \mathrm{e}^-,$$

$$\nu_x + \mathrm{d} \rightarrow \mathrm{p} + \mathrm{n} + \nu_x,$$

$$\nu_x + \mathrm{e}^- \rightarrow \nu_x + \mathrm{e}^-,$$

where the subscript x denotes any type of flavor. Only electron neutrinos produce charged-current interactions, while the neutral-current (NC) interactions and elastic scattering are sensitive to all types of flavors. The NC reaction measures the total flux of all active neutrino flavors produced in the Sun with energy above the threshold of 2.2 MeV. This provides a measurement of the solar neutrino flux as

$$5.21 \pm 0.27\,(\text{statistical uncertainty}) \pm 0.38\,(\text{systematic uncertainty}) \times 10^6\,\text{cm}^{-2}\,\text{s}^{-1},$$

which is in agreement with the standard solar models. A global analysis of these and other solar- and reactor-neutrino results yields $\Delta m^2 = 7.1\,^{+1.2}_{-0.6} \times 10^{-5}\,\text{eV}^2$ and $\theta = 32.5\,^{+2.4}_{-2.3}$ degrees.

- *Accelerator neutrinos* The LSND experiment has found an electron-antineutrino excess in a muon-antineutrino beam. The solution would be small mixing and a mass-squared difference of 0.1–1 eV^2. However, the very similar KARMEN experiment has found no such effect, but it cannot rule out LSND's complete parameter space. The LSND results are referred to as the LSND anomaly because the large mass difference cannot be reconciled with the three-family model. The MiniBoone experiment at Fermilab is currently running with the aim of checking the LSND result.
- *Long-base-line experiments* There are several experiments that will use two detectors, one close to the accelerator, where the neutrino beam is produced, and a second detector at a distance of 200–400 km. The nearby detector will be used for calibration reasons to determine the beam and properties of the neutrino reactions. The faraway detector will be looking at the changes that take place because muon neutrinos oscillate to neutrinos of another flavor. A small-scale experiment, K2K, operated in Japan and two others are under construction, MINOS in the USA and OPERA in Europe.

For three flavors of neutrinos there are only two independent Δm^2. If the result from the LSND experiment is correct, then there must be a fourth flavor, which has to be "sterile," i.e. it does not couple to the gauge bosons and therefore does not contribute to the Z^0 width. The results of the other experiments are summarized as follows:

$$(\Delta m^2\,(\text{eV}^2),\ \sin^2\theta) \simeq \begin{cases} (2 \times 10^{-3}, \simeq 0.5), & \text{atmospheric,} \\ (7 \times 10^{-5}, \simeq 0.3)\ \text{LAMSW}, & \text{solar, KamLAND.} \end{cases} \tag{13.36}$$

We can now estimate the magnitude of the leptonic MNS matrix. Let us ignore the LSND result and assume that there are two mass-squared differences governing the atmospheric and solar oscillation. They obey $\Delta m^2_{\odot} \ll \Delta m^2_{\mathrm{A}}$. If we identify $\Delta m^2_{21} = \Delta m^2_{\odot} \ll \Delta m^2_{\mathrm{A}} = \Delta m^2_{31} \simeq \Delta m^2_{32}$, then we have, for a short-baseline reactor experiment such as CHOOZ (see Problem 2),

$$P^{\text{CHOOZ}}_{\text{ee}} = 1 - 4|U_{\text{e}3}|^2(1 - |U_{\text{e}3}|^2)\sin^2\Delta_{31}, \tag{13.37}$$

where $\Delta_{ij} = \Delta m_{ij}^2 \, L/(4E)$ and L is the distance for the CHOOZ experiment. The absence of disappearance of neutrinos in the experiment means that $|U_{e3}|$ is either small or close to unity. The probability for solar neutrinos is (Bilenky, 2003)

$$P_{ee}^{\odot} = (1 - |U_{e3}|^2)^2 \left(1 - 4\frac{|U_{e1}|^2 |U_{e2}|^2}{(1 - |U_{e3}|^2)^2} \sin^2\Delta_{21} \right) + |U_{e3}|^4. \tag{13.38}$$

Experimentally, $P_{ee}^{\odot}$ is significantly less than unity and also energy-dependent. On combining the results of the CHOOZ experiment with the solar deficit, we conclude that $|U_{e3}|^2 \ll 1$. Finally, atmospheric neutrinos oscillate with

$$P_{\mu\tau}^{A} = 4|U_{\mu 3}|^2 |U_{\tau 3}|^2 \sin^2\Delta_{31}. \tag{13.39}$$

Note that oscillations of reactor and atmospheric neutrinos are triggered by the same Δm^2. If for the latter we use $\Delta m^2 = 3 \times 10^{-3}$ eV^2, CHOOZ gives $|U_{e3}|^2 \lesssim 0.05$. As a first approximation we can assume $|U_{e3}| \simeq 0$, maximal atmospheric mixing, and the LAMSW solution for solar mixing. With these results, we can approximate the MNS matrix by

$$U_{\alpha i} \simeq \begin{pmatrix} c & s & 0 \\ -s/2 & c/2 & 1/\sqrt{2} \\ s/2 & -c/2 & 1/\sqrt{2} \end{pmatrix} \text{LAMSW}, \tag{13.40}$$

where $c = \cos\theta$, $s = \sin\theta$, and $s^2 \simeq 0.3$. Note that there can be a phase in the matrix, resulting in CP violation as in the quark sector. However, for $U_{e3} = 0$ the theory is effectively a two-flavor theory. Only for a non-vanishing U_{e3} element could one establish CP violation in long-baseline experiments.

If neutrinos are Majorana particles (see below), then there are two additional phases in the mixing matrix. They can be shown to have no influence on oscillation physics and reveal their presence only in neutrinoless double beta decay, which we will discuss below.

13.4 Majorana neutrinos

The results of the last section give strong evidence in favor of massive neutrinos. The oscillations indicate that in the leptonic sector flavor number is not conserved, i.e. muon neutrinos can become tau neutrinos, etc. This is a change from one family to the next. So far there has been no discussion of a particle changing into its antiparticle. The mixing among the families is produced when we introduce a mass term of the form

$$\mathcal{L}^{D} = \sum_{i,j} m_D^{ij} \overline{\psi}_i \psi_j + \text{h.c.}, \tag{13.41}$$

with i and j running over the families. For the electroweak theory, special attention is required because the states are classified according to helicities – with left-handed particles in doublets and right-handed particles in singlets of weak SU(2). The mass term is now

$$\mathcal{L}^{\mathrm{D}} = \sum_{i,j} m_{\mathrm{D}}^{ij} \bar{\psi}_{\mathrm{L}i} \psi_{\mathrm{R}j} + \text{h.c.} \tag{13.42}$$

It is known as the Dirac mass term and is produced by the Higgs mechanism, as described at the beginning of this chapter.

There is also the possibility of introducing new mass terms,

$$\frac{1}{2} M_{\mathrm{R}} \bar{N}_{\mathrm{R}}^{\mathrm{c}} N_{\mathrm{R}} \quad \text{and} \quad \frac{1}{2} M_{\mathrm{L}} \bar{\nu}_{\mathrm{L}}^{\mathrm{c}} \nu_{\mathrm{L}} \tag{13.43}$$

and their Hermitian conjugates, which are called Majorana mass terms. Obviously, Majorana mass terms can include mixing among the generations by introducing indices i and j, as in Eq. (13.42). They also mix neutrinos with antineutrinos. The new terms are Lorentz-invariant, and they carry lepton number two, thus introducing violation of the lepton number. Again, special attention must be paid to the peculiar property of the electroweak theory which classifies the states according to helicities. This introduces special requirements on the terms which are allowed. For instance, the Lorentz structure gives the identities $\bar{N}_{\mathrm{R}} N_{\mathrm{R}} = \bar{\nu}_{\mathrm{L}} \nu_{\mathrm{L}} = 0$ and $\bar{N}_{\mathrm{R}} \nu_{\mathrm{L}}^{\mathrm{c}} = \bar{\nu}_{\mathrm{L}} N_{\mathrm{R}}^{\mathrm{c}} = 0$. A Majorana state is defined as one with equal components of particles and antiparticles. It follows from the above identities that

$$\left(\bar{N}_{\mathrm{R}}^{\mathrm{c}} + \bar{N}_{\mathrm{R}}\right)\left(N_{\mathrm{R}}^{\mathrm{c}} + N_{\mathrm{R}}\right) = \bar{N}_{\mathrm{R}}^{\mathrm{c}} N_{\mathrm{R}} + \bar{N}_{\mathrm{R}} N_{\mathrm{R}}^{\mathrm{c}},$$

which indicates that the mass terms introduced do indeed correspond to Majorana particles.

Similarly, specific terms are permitted by the SU(2) symmetry. We can introduce the term $M_{\mathrm{R}} \bar{N}_{\mathrm{R}}^{\mathrm{c}} N_{\mathrm{R}}$, since it is Lorentz-invariant and SU(2) singlet. We cannot introduce the term $M_{\mathrm{L}} \bar{\nu}_{\mathrm{L}}^{\mathrm{c}} \nu_{\mathrm{L}}$, because it is the direct product of a doublet with a doublet, which decomposes into an SU(2) singlet plus a triplet. Their product with the Higgs doublet cannot produce a singlet. Similarly, their product with a singlet produces a singlet and a triplet. SU(2) symmetry dictates that $M_{\mathrm{L}} = 0$, unless we introduce triplet representations of Higgses. The possible mass terms are a Dirac mass term as in Eq. (13.42) and, in addition, a Majorana term appearing as the first term in Eq. (13.43). We sum up these terms in a matrix notation:

$$\mathcal{L}_{\text{mass}} = \left(\bar{\nu}_{\mathrm{L}} \bar{N}_{\mathrm{R}}^{\mathrm{c}}\right) \begin{pmatrix} 0 & m_{\mathrm{D}} \\ m_{\mathrm{D}} & M_{\mathrm{R}} \end{pmatrix} \begin{pmatrix} \nu_{\mathrm{L}}^{\mathrm{c}} \\ N_{\mathrm{R}} \end{pmatrix} + \text{h.c.} \tag{13.44}$$

We can diagonalize this mass matrix with an orthogonal transformation. The eigenvalues are

$$\lambda_{1,2} = \frac{1}{2}\left(M_{\mathrm{R}} \pm \sqrt{M_{\mathrm{R}}^2 + 4m_{\mathrm{D}}^2}\right), \tag{13.45}$$

which for $m_{\mathrm{D}}/M_{\mathrm{R}} \ll 1$ become

$$\lambda_1 = -\frac{m_{\mathrm{D}}^2}{M_{\mathrm{R}}} \quad \text{and} \quad \lambda_2 = M_{\mathrm{R}} + \frac{m_{\mathrm{D}}^2}{M_{\mathrm{R}}}. \tag{13.46}$$

The case we studied is called the seesaw mechanism (Gell-Mann *et al.*, 1979; Yanagida, 1979; Minkowski, 1977) because when one mass is large the other is small. Now, since the neutrino masses are very small relative to lepton and quark masses, for instance $m_\nu \approx 10^{-2}$ eV, the seesaw mechanism supplies an explanation provided that the Majorana mass, M_{R}, is large.

The corresponding wave functions, omitting a normalization constant, are

$$\begin{aligned} \psi_1 &= \nu_{\mathrm{L}} + \frac{m}{M_{\mathrm{R}}} N_{\mathrm{R}}^{\mathrm{c}}, \\ \psi_2 &= -\frac{m}{M}\nu_{\mathrm{L}}^{\mathrm{c}} + N_{\mathrm{R}}, \end{aligned} \tag{13.47}$$

indicating that ψ_1 consists primarily of the normal neutrino with a small admixture of $N_{\mathrm{R}}^{\mathrm{c}}$. The other state ψ_2 is the heavy state with a large component N_{R} and a small admixture of $\nu_{\mathrm{L}}^{\mathrm{c}}$.

Identifying ν_{L} or ψ_1 with the standard-model neutrinos explains their lightness by introducing a heavy scale. It follows a hierarchical mass scheme for the neutrinos. Neutrino masses $m_\nu \simeq 10^{-2}$ eV require $m_{\mathrm{R}} \simeq 10^{16}$ GeV for a Dirac scale of 1 GeV. This is a typical scale of grand unified theories, which is one of the reasons why seesaw models are very popular.

13.5 Neutrinoless double beta decay

Much effort has been devoted to discovering the nature of the neutrino. The experimental results on oscillations are independent of the Majorana character of the neutrinos. They observe oscillation of flavors. Experiments also search for evidence of Majorana neutrinos. A reaction that is conceptually simple concerns the conversion

$$\nu_{\mathrm{e}} + \mathrm{N} \to \mathrm{e}^+ + \text{hadrons},$$

which has, unfortunately, a very small cross section. Theoretical estimates indicate that the rate for a neutrino with an energy of 1 GeV is 10^{-18}–10^{-22} times smaller

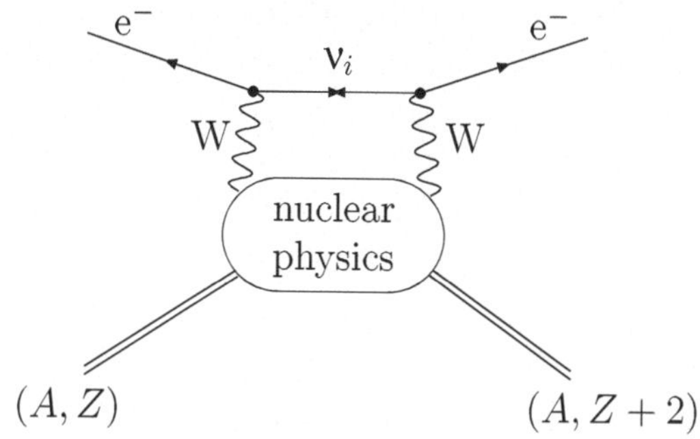

Figure 13.1. Neutrinoless double beta decay.

than the normal charged-current cross section. Experiments with neutrino beams are far away from these limits.

A favorable reaction is

$$2\text{n} \to 2\text{p} + (\text{e}^-\text{e}^-),$$

which is known as neutrinoless double beta decay and is indicated as $0\nu\beta\beta$. The process is shown in Fig. 13.1 where a nucleus with Z protons and $A-Z$ neutrons emits two W^- bosons and decays into a nucleus with $Z+2$ protons and $A-Z-2$ neutrons. The two W^- bosons interact with each other and convert into two electrons. The lepton propagator is a Majorana neutrino containing both particles and antiparticles. The amplitude for the process contains the leptonic tensor

$$L_{\alpha\beta}(0\nu\beta\beta) \propto U_{\text{ei}}^2 \bar{e}\gamma_\alpha\gamma_- \frac{\not{p}+m_i}{p^2-m_i^2}\gamma_\beta\gamma_+ e = U_{\text{ei}}^2 m_i \frac{1}{p^2}\bar{e}\gamma_\alpha\gamma_-\gamma_\beta e, \quad (13.48)$$

where p is the momentum of the propagating neutrino, m_i its mass, and e are the electron fields. The element U_{ei} gives the coupling of the electron to the ith neutrino mass state. There is also a corresponding hadronic tensor $W^{\alpha\beta}$, whose structure is rather complicated. Its structure is unfortunately left out in most theoretical articles and discussions in spite of the fact that it introduces considerable uncertainty in the predictions. The decay rate of the process is obtained by first summing the amplitudes over all intermediate states and then squaring the total amplitude to arrive at

$$\Gamma(0\nu\beta\beta) = A\left|\sum_i U_{\text{ei}}^2 m_i\right|^2, \quad (13.49)$$

where A is a non-trivial factor representing nuclear matrix elements. The quantity

$$m_{\text{ee}} = \sum_i U_{\text{ei}}^2 m_i, \quad (13.50)$$

with the sum running over all neutrinos, is called the effective electron-neutrino Majorana mass, or in short the effective neutrino mass. The neutrino masses in the propagator are physical masses and are very small relative to q^2 involved in nuclear decays. The masses m_i and the mixing elements U_{ei} also appear in the oscillations of neutrinos. Unfortunately, the oscillation experiments measure Δm_{ij}^2, i.e. mass differences, rather than absolute values for the masses. They also measure the squares of the U_{ei}, which in general are complex functions. Thus, extracting mass differences and mixing angles from oscillation experiments, there is still a large range of values possible for m_{ee} (Pascoli *et al.*, 2005; Choubey and Rodejohann, 2005). Processes depending on the effective Majorana mass have branching ratios and cross sections that can be very small. The current limit for m_{ee} comes from neutrinoless double beta decay of ^{76}Ge and is approximately 0.2 eV. There is only one experiment (Klapdor-Kleingrothaus *et al.*, 2004) in which it has been claimed that the Majorana mass has been measured. The best value given by the group is $\langle m_{ee} \rangle = 0.39$ eV, but the data analysis of this experiment has been criticized (Aalseth *etal.*, 2004). There are plans to build improved experiments. This is a difficult but very interesting and exciting field, regarding which the reader can consult other articles (Aalseth *et al.*, 2004).

Problems for Chapter 13

1. Show that the mixing angle θ in the matrix

$$U = \begin{pmatrix} \cos\theta & \sin\theta \\ -\sin\theta & \cos\theta \end{pmatrix}$$

which diagonalizes the symmetric matrix

$$M = \begin{pmatrix} a & b \\ b & d \end{pmatrix}$$

is given by

$$\tan(2\theta) = \frac{2b}{d-a}$$

and the eigenvalues

$$E_{1,2} = \frac{1}{2}\left[(a+d) \pm \frac{2b}{\sin(2\theta)}\right].$$

What happens to the mixing angle and the difference of the eigenvalues of M when one adds a term proportional to the unit matrix to M?

2. For three generations there are only two independent mass differences. For this reason the observation of two oscillations, $e \to \mu$ and $\mu \to \tau$, determines the mass-squared differences

$$\Delta m_{\odot}^2 = \Delta m_{12}^2,$$
$$\Delta m_{\mathrm{A}}^2 = \Delta m_{23}^2 = \Delta m_{31}^2.$$

For the CHOOZ experiment the distance is approximately a kilometer, so $\Delta m_{12}^2 \, L/E$ is very small and does not contribute to the oscillation.

(i) Use Eq. (13.22) and approximations suggested by the description of this problem to derive Eq. (13.39).

(ii) Use again Eq. (13.22), the same approximations, and the unitarity of the mixing matrix to derive Eq. (13.37).

3. The general Lagrangian with Dirac and Majorana mass terms is

$$-L = \bar{\psi}\, \partial\!\!\!/ \psi + M_\mathrm{D}\left[\bar{\psi}_\mathrm{L}\psi_\mathrm{R} + \mathrm{h.c.}\right] + \frac{M_\mathrm{L}}{2}\left[(\bar{\psi}_\mathrm{L})^\mathrm{c}\psi_\mathrm{L} + \mathrm{h.c.}\right] + \frac{M_\mathrm{R}}{2}\left[(\bar{\psi}_\mathrm{R})^\mathrm{c}\psi_\mathrm{R} + \mathrm{h.c.}\right],$$

with M_L and M_R being Majorana mass terms for the left- and right-handed neutrinos.

(i) Show that it can be written in the form

$$-L = \bar{V}\, \partial\!\!\!/ V + \bar{V}[M]V,$$

where V is a column matrix,

$$V = \frac{1}{2}\begin{pmatrix} \psi_\mathrm{L} + (\psi_\mathrm{L})^\mathrm{c} \\ \psi_\mathrm{R} + (\psi_\mathrm{R})^\mathrm{c} \end{pmatrix},$$

and $[M]$ is a symmetric matrix,

$$[M] = \begin{pmatrix} M_\mathrm{L} & M_\mathrm{D} \\ M_\mathrm{D} & M_\mathrm{R} \end{pmatrix}.$$

$[M]$ is the neutrino mass matrix. For $M_\mathrm{L} = 0$ it reduces to the seesaw case.

(ii) Let ψ_1 and ψ_2 be the eigenvector fields of the mass matrix with eigenvalues λ_1 and λ_2, respectively. Then show that L can be rewritten as

$$-L = \bar{\psi}_1\, \partial\!\!\!/ \psi_1 + \bar{\psi}_2\, \partial\!\!\!/ \psi_2 + \lambda_1\bar{\psi}_1\psi_1 + \lambda_2\bar{\psi}_2\psi_2.$$

The physics content of the Lagrangian is now clear: it is the free Lagrangian for two particles ψ_1 and ψ_2.

4. Consider the Dirac Lagrangian and add the term

$$\bar{\psi}(x)\gamma_5\psi(x)\phi.$$

The Dirac equation now becomes

$$[\mathrm{i}\,\partial\!\!\!/ - \gamma_5\phi - m]\psi(x) = 0.$$

When $\phi(x)$ is a function of x_μ, the addition term represents the interaction of the fermion with a scalar field. We may also consider ϕ to be a constant, i.e. independent of space and time. Consider the latter case and search for plane-wave solutions

$$\psi(x) = u(p)\mathrm{e}^{-\mathrm{i}p\cdot x}.$$

(a) Which condition must p_μ satisfy in order for solutions to exist?

(b) Once p_μ satisfies the conditions, which are the linearly independent spinors?

(c) Find a chiral transformation $T = \mathrm{e}^{\mathrm{i}\gamma_5\theta}$ so that the Dirac equation is brought into the form $(\mathrm{i}\,\partial\!\!\!/ - \tilde{m})\psi(x) = 0$.

Comment This exercise shows that, whenever ϕ is a constant, the term $\phi\bar{\psi}(x)\gamma_5\psi(x)$ is a mass term. It also follows that the sign of the fermion mass can be changed by a chiral transformation.

References

Aalseth, C., Back, H., Dauwe, L., *et al.* (2004), hep-ph/0412300
Bilenky, S. M. (2003), hep-ph/0210128, Eq. (83)
Choubey, S., and Rodejohann, W. (2005), *Phys. Rev.* **D72**, 033016
Gell-Mann, M., Ramond, P., and Slansky, R. (1979), "Complex spinors and unified theories," in *Supergravity*, ed. P. van Niewenhuisen and D. Z. Freedman (Amsterdam, North-Holland), p. 315
Klapdor-Kleingrothaus, H. V., Krivosheina, J. V., Dietz, A., and Chkvoret, O. (2004), *Phys. Lett.* **B586**, 198
Maki, Z., Nakagawa, M., and Sakata, S. (1962), *Prog. Theor. Phys.* **28**, 870
Mikheyev, S., and Smirnov, A. (1985), *Yad. Fiz.* **42**, 4441
Minkowski, P. (1977), *Phys. Lett.* **B67**, 421
Mohapatra, R. N., Antusch, S., Babu, K. S. *et al.* (2005), APS Study, hep-ph/0510213
Pascoli, S., Petcov, S., and Schwetz, T. (2005), hep-ph/0505226
Wolfenstein, L. (1978), *Phys. Rev.* **D17**, 2369
Yanagida, T. (1979), "Horizontal symmetries and masses of neutrions," in *Proceedings of the Workshop on the Unified Theory and Baryon Number in the Universe*, ed. O. Sawada and A. Sugamoto (Tsukuba, Japan, KEK), p. 95

Select bibliography

Bjorken, J. D., and Drell, S. D. (1965), *Relativistic Quantum Mechanics* (New York, McGraw-Hill)
Kayser, B. (2000), Neutrino mass, mixing and flavor change, in *Review of Particle Physics*, *Eur. Phys. J.* **C15**, 344
Mohapatra, R. N., and Pal, P. B. (1991), *Massive Neutrinos in Physics and Astrophysics* (Singapore, World Scientific)

14

Heavy quarks

14.1 Introduction

When quarks were introduced into physics, they were considered to be light, like the up, the down, and the strange quarks. Their bound states were light mesons, such as the π and K mesons. In fact, light states were interpreted as the Goldstone bosons of the SU(3) symmetry with several of the aspects of Goldstone particles discussed in Chapter 5.

The next quark is the charm quark, which was formulated in order to suppress flavor-changing-neutral couplings in the K mesons (Glashow *et al.*, 1970) (GIM henceforth). Since the charm quark is much heavier than the proton, its existence gave rise to the possibility that additional heavy quarks may exist. The expectation was confirmed with the discoveries of the bottom and top quarks when accelerators of higher and higher energies began operating.

The precise definition of quark masses is a delicate topic and for this reason we shall discuss some of the issues involved. Masses of fermions appear in the electroweak Lagrangian after the breaking of the symmetry, i.e. when the Higgs field acquires a vacuum expectation value. Masses for particles are measured through their interactions with an external field; for example, the bending of an electron beam in a magnetic field determines the ratio e/m. The interaction contains higher-order corrections, which must be included. For leptons the masses are defined as poles of the propagators. For quarks the situation is more complicated because they never appear as free particles, but are always confined within hadrons. The masses of quarks must include radiative corrections from the forces which confine them. On the energy scale of the heavy quarks the strong coupling constant is small enough to allow perturbative calculations. We define the running quark mass $m(p^2)$ as the renormalized mass parameter in the quark propagator

$$S(p) = \frac{\mathrm{i}}{\not{p} - m_0 - \Sigma(p) + \mathrm{i}\varepsilon} = \frac{\mathrm{i}Z_3(p)}{\not{p} - m(p^2) + \mathrm{i}\varepsilon}. \tag{14.1}$$

In this expression m_0 is the bare mass and $\Sigma(p)$ is the fermion self-energy correction obtained in perturbation theory, which contains infinite terms. The infinite terms combine with m_0 to define the physical mass and define the wave-function renormalization $Z_3(p)$. The specific method for eliminating (subtracting) the infinity is scheme-dependent and brings into the definition of masses and coupling constants the renormalization scale μ_0, where the subtraction takes place. This is very similar to the definition of the running coupling constant described in Section 11.2.

So far have we defined quark masses as functions of the renormalization point μ_0 and the momentum p at which they are measured. We need an additional prescription for relating them to masses of hadrons or for describing how we estimate reactions in which heavy quarks occur. There are several methods for extracting quark masses from physical processes. One of them uses sum rules over two-point functions (spectral functions), which are saturated by physical data and are matched to theoretical expressions that include running quark masses (Narison, 2001; Manohar, 2000). Another method appeals to decays of mesons containing a single heavy quark. Here one computes the decay in a two-fold expansion: in powers of $\alpha_s(p)$ and a non-perturbative series in powers of $\Lambda_{\mathrm{QCD}}/m_Q$ (Manohar and Wise, 2000). The masses extracted by the various methods are close to each other and contain an explicit dependence on the momentum at which they are measured. The spectrum that emerges for the new quarks is

$$m_{\mathrm{c}}(m_{\mathrm{c}}) = 1.15\text{–}1.35\,\mathrm{GeV},$$
$$m_{\mathrm{b}}(m_{\mathrm{b}}) = 4.6\text{–}4.9\,\mathrm{GeV}.$$

The mass of the top quark is determined in another way. The observed events are attributed to the production of $\mathrm{t\bar{t}}$ pairs and their subsequent decays into leptons and hadron jets. The production of the pairs is computed in the parton model using lowest-order QCD. The value for the top quark appears in the calculation and the value which optimizes the fit is reported as the standard-model mass, giving (CDF/DO, 2005)

$$m_{\mathrm{t}} = 174.3 \pm 3.4\ \mathrm{GeV}.$$

Besides the masses of the heavy quarks we need their couplings to charged and neutral currents. They are accurately determined by the gauge nature of the theory and experimental data. The couplings have already been discussed in Chapter 9.

There are several other properties of heavy quarks that we take up in this chapter. First, the decays of states containing heavy quarks involve hadronic matrix elements that simplify considerably. In addition, when they appear as intermediate states in loop diagrams, they dominate the diagrams and produce new phenomena such as the mixing of states and CP asymmetries. Examples of the new phenomena are (i) B^0–$\bar{B}^0$ mixing dominated by box diagrams and (ii) some CP asymmetries dominated by penguin diagrams. Finally, heavy quarks may have strong couplings to induce bound states that imitate the Higgs particles. Some of these topics will be covered in this and the following chapters.

It is still interesting to ask which quarks are considered to be heavy. A heavy state is one with

$$m_q \gg \Lambda_{QCD},$$

where according to QCD the interparticle forces become weak. For instance, the inclusive semileptonic decays of pseudoscalar mesons containing a heavy quark must be given to a first approximation by the spectator model. The spectator model, improved by the momentum distribution of a b quark in a B meson, gives an accurate description of semileptonic B-meson decays. Thus B and higher meson states are considered to be heavy states. In fact, for hadronic states containing a heavy quark, a systematic expansion has been developed in inverse powers of the heavy-quark mass, which will be described in Sections 14.4 and 14.5. The top quark decays so fast that bound states do not have enough time to form.

A crude test for the validity of the spectator model is provided by the ratio of lifetimes of charged and neutral mesons. For mesons with beauty quarks,

$$\frac{\tau_{B^+}}{\tau_{B^0}} = 1.08 \pm 0.01,$$

which is to be compared with the ratio for mesons containing charmed quarks,

$$\frac{\tau_{D^+}}{\tau_{D^0}} = 2.55 \pm 0.01.$$

The ratio for the B mesons suggests that the spectator model is applicable. The ratio for D mesons indicates that there are additional contributions. In fact, it was a surprise when the experimental colleagues established this large ratio. It is now generally believed that annihilation diagrams are important in D-meson decays. The simple picture that mesons consist of a quark and an antiquark is too naive, because the gluons in D mesons play a very active role in binding the quarks. We illustrate this fact by the following argument, where we set the CKM matrix equal to the unit matrix, which makes the V_{cd} element zero. The spectator decay mode (see Section 14.2)

$$c \to s + \bar{d} + u$$

is expected to give equal contributions to D^0 and D^+ decays, and its amplitude will be denoted by A_{sp}. In addition, for the D^0 meson there is the annihilation diagram for the reaction

$$c + \bar{u} \to s + \bar{d} + \text{gluons},$$

with a W^+ exchanged in the t-channel. Its amplitude will be denoted by A_{an}. The decay width for D^0 depends on the sum of the two amplitudes,

$$\Gamma_{D^0} \propto |A_{sp}|^2 + |A_{an}|^2,$$

and similarly for D^+,

$$\Gamma_{D^+} \propto |A_{sp}|^2.$$

The CKM-matrix elements are the same in all these diagrams, indicating that the lifetime τ_{D^+} is larger than τ_{D^0}. Precise calculations for these and other decays of charm mesons are difficult, but there are phenomenological models that give a consistent picture for several decay channels (Bander *et al.*, 1980).

This chapter is long and contains various topics. To help the reader I outline its contents. The following three sections describe decays of heavy quarks in phenomenological models. Some of these models have been very popular. Sections 14.4 and 14.5 are devoted to the heavy-quark effective theory (HQET), which is a systematic expansion of amplitudes in inverse powers of the heavy-quark mass. The top quark is very heavy and has special properties. For this reason a special section is devoted to it. Finally, heavy quarks play an important role in loop diagrams, with a simple example being provided by box diagrams, which are ultraviolet-finite. An introduction to the computational methods for box diagrams is included in Section 14.7.

14.2 Semileptonic and inclusive B-meson decays

14.2.1 The spectator model

When a heavy quark, generically denoted by Q, decays inside a hadron, it does so without disturbing the surrounding field produced by other quarks, antiquarks, and gluons. In the B mesons, for instance, the b quark decays, leaving the light antiquark u, d, or s (hereafter generically denoted as q) undisturbed (spectators). Thus we expect the decay of the B meson to be given by the free decay of the b quark to a zeroth-order approximation. This result must be improved for the bound-state effects of the meson, as will be described in this section (Altarelli *et al.*, 1982).

In the spectator model, the light quark moves in the field of the heavy quark, and between them they share all the energy and momentum of the meson. The spectator quark q has a definite mass m and a four-momentum $p_q \equiv (E_q, \mathbf{p})$. In the rest frame of the meson the b quark moves with a virtual mass W and a

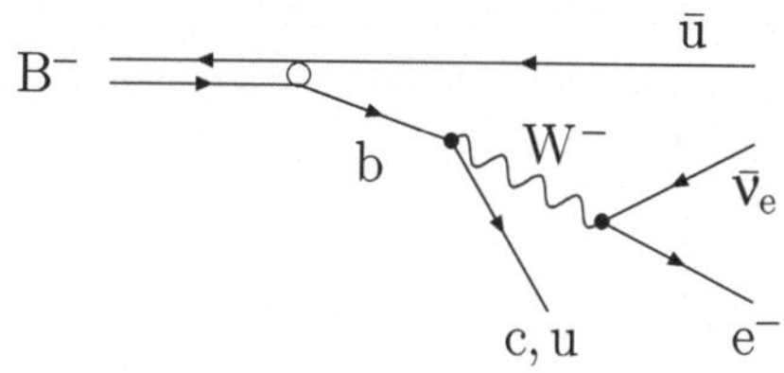

Figure 14.1. B-meson decay.

four-momentum $p_b \equiv (E_b, -\mathbf{p})$, where

$$\begin{aligned} E_q + E_b &= M_B, \\ W^2 &= M_B^2 + m_q^2 - 2M_B\sqrt{\mathbf{p}^2 + m_q^2}, \end{aligned} \tag{14.2}$$

and M_B is the mass of the B meson. The decay is now given by the point-like decay of the b quark averaged over its momentum inside the meson (Fig. 14.1).

In the model of Altarelli *et al.* (1982), henceforth called the ACCMM model, the averaging over the momentum of the b quark is done by introducing a Gaussian distribution function,

$$\phi(|\mathbf{p}|) = \frac{4}{\sqrt{\pi}\, p_F^3} \exp\big(-|\mathbf{p}|^2/p_F^2\big), \tag{14.3}$$

where p_F, the Fermi momentum, is a free parameter to be adjusted by comparing the theoretical prediction with the experimental spectrum. Its value is of the order of a few hundred MeV at most, so large-$\mathbf{p}$ configurations are exponentially suppressed. With this distribution the b quark remains close to its mass shell.

To calculate the electron spectrum from the semileptonic decay $b \to u(c)e\bar{\nu}_e$, one follows three steps.

1. One calculates the differential decay width for a point quark in its rest frame. This is exactly analogous to the muon decay and the result is

$$\frac{d\Gamma^0}{dE} = \frac{G^2 W^4}{48\pi^3} |V_{u(c)b}|^2 x^2 (3 - 2x). \tag{14.4}$$

Here, E is the electron energy, V_{ub} or V_{cb} are the CKM-matrix elements, and $x = 2E/W$. Note that we have neglected the masses of all final-state particles, which is surely not justified for a $b \to c$ transition, with the relevant mass dependences given in the ACCMM article.

2. In addition, one may incorporate the QCD corrections to this tree-level decay distribution, by multiplying $d\Gamma^0/dE$ by an overall factor that is less than unity and depends upon x. This changes the shape of the spectrum near its endpoint. This will not be discussed here, but is described briefly in Problem 2.

3. The decaying quark is not at rest but moves with a momentum $|\vec{p}|$ inside the meson. We can account for this motion by writing the decay rate in a Lorentz-invariant form, which will be valid in any frame. To this end, we write first

$$x = \frac{2E}{W} = \frac{2p_b \cdot p_e}{W^2} = \frac{2}{W^2}(E_b E_e - p p_e \cos\theta). \tag{14.5}$$

The first two equations are computed in the rest frame of the b quark and the third one in a frame where the b quark moves. Here $E_b^2 = W^2 + p^2$, $p_e^2 = E_e^2 - m_e^2$, and θ is the angle between the momentum of the b quark and the direction of the electron. The final decay rate is obtained by averaging over p and θ using the distribution function

$$\frac{d\Gamma}{dE_e} = \frac{G^2}{48\pi^3}|V_{ub}|^2 \int_0^{p_F} W^4 x^2 (3 - 2x)\phi(p) p^2 \, dp \, d\cos\theta; \tag{14.6}$$

here x depends on p and the range of integration over θ is from 0 to π.

This method of averaging over the Fermi motion respects Lorentz invariance and takes into account the phase space. It is somewhat crude, in the sense that the energy and momentum of the meson are supposed to be distributed between the two constituent quarks only and the distribution formula is valid only when $|\mathbf{p}|$ is rather small. In spite of its crudeness, the model describes accurately the electron spectrum observed in the experiments. It indicates that for B and heavier mesons the spectator model is a reasonable zeroth-order approximation, requiring additional corrections from the interaction of the heavy quark with its surrounding field. We shall discuss some improvements in this chapter.

14.2.2 The parton model

The semileptonic B-meson decays can be presented by the diagram in Fig. 14.2, where X can be a specific final state or the incoherent sum over various final states. The drawing represents the square of the amplitude and looks very much like the diagram for deep inelastic scattering. This suggests that a formally similar analysis for the B-meson inclusive decays is possible. There are two differences, however; the initial B meson is heavy and the current is time-like, with its variables determined by the initial and final hadronic states. Consequently, we use the same formalism but the structure function is now different, giving the probability of finding a b quark in a fast-moving B meson.

In a standard analysis the square of the matrix element is given as the product of a leptonic and a hadronic tensor,

$$|\mathcal{M}|^2 = L^{\mu\nu} W_{\mu\nu}. \tag{14.7}$$

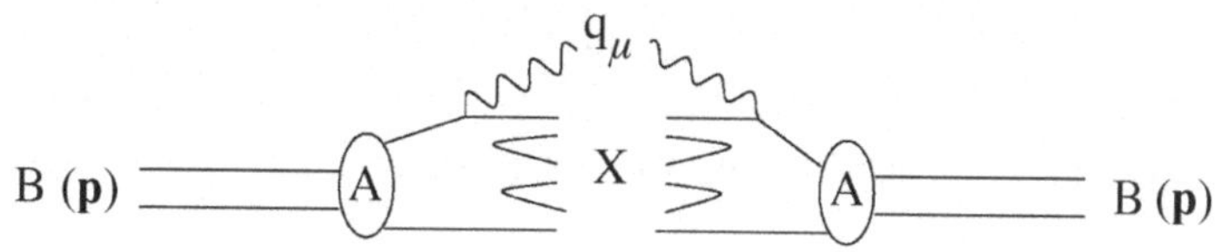

Figure 14.2. Semileptonic B decay in the parton model.

The hadronic tensor can be written as

$$W_{\mu\nu}(q^2, q\cdot p) = \frac{1}{2\pi}\int d^4x\, e^{iqy}\langle B|[J_\mu(x), J_\nu(0)]|B\rangle$$
$$= -g_{\mu\nu}W_1 + p_\mu p_\nu W_2 + i\varepsilon_{\mu\nu\alpha\beta}p^\alpha q^\beta W_3 + \cdots, \qquad (14.8)$$

with the notation being analogous to that introduced for neutrino reactions. After analyzing the kinematics of the decay, it was shown that a large region of phase space is dominated by the singular behavior of the commutator on the light cone. This allows us to express the decay in terms of a quark distribution function $f(x)$, which denotes the probability of finding a b quark in the B meson. Dominance of the light cone demands that the analysis of the decay be carried out in an infinite-momentum frame.

We visualize the B meson moving in an infinite-momentum frame, where the motion of the constituents slows down relative to the time of the decay. In this frame the b quark decays, leaving the rest of the B meson undisturbed. The decay is in general given by

$$\frac{d\Gamma}{dE_e\, dq_0\, dq^2} = \frac{G^2|V_{ub}|^2}{32\pi^3 m_B}\left\{2q^2W_1 + \left(4E_eq_0 - 4E_e^2 - q^2\right)W_2 - 2q^2(2E_e - q_0)\frac{W_3}{m_B}\right\}. \qquad (14.9)$$

The hadronic tensor for point-like interaction is

$$W_0^{\mu\nu}(q^2, q\cdot p) = \frac{1}{2}\,\mathrm{Tr}\left[\not{p}_u\gamma^\mu(1-\gamma_5)\not{p}_b\gamma^\nu(1-\gamma_5)\right]\delta\left[(p_b - q)^2\right], \qquad (14.10)$$

with p_u and p_b the four-momenta of the u and b quarks, respectively. The b quark carries a fraction x of the B meson's momentum and has the momentum distribution function $f(x)$. We arrive at the B-meson decay by substituting

$$p_b = xP \quad \text{and} \quad p_u = p_b - q, \qquad (14.11)$$

then multiplying $W_0^{\mu\nu}$ by $f(x)$ and integrating over x,

$$W^{\mu\nu} = \int W_0^{\mu\nu}(q^2, xq\cdot p)f(x)\delta(x^2p^2 - 2xq\cdot p + q^2)dx. \qquad (14.12)$$

The δ-function has two roots $x_\pm = (q_0 \pm |\vec{q}|)/M_B$. The smaller root x_- corresponds to diagrams with particles moving backwards in time and will be neglected because it is numerically small. The structure functions are given as (Bareiss and Paschos, 1989)

$$W_1 = 2x_+ f(x_+), \qquad W_2 = \frac{4M_B}{|\vec{q}|} x_+^2 f(x_+), \qquad W_3 = -\frac{2M_B}{|\vec{q}|} x_+ f(x_+). \tag{14.13}$$

Direct substitution leads to the final result

$$\frac{d\Gamma}{dE_e\, dq_0\, dq^2} = \frac{G^2|V_{ub}|^2}{4\pi^3 m_B} \frac{xf(x)}{\sqrt{q_0^2 - q^2}} (q_0 - E_e)\big(2E_e m_B x - q^2\big), \tag{14.14}$$

with $x = (q_0 + |\vec{q}|)/M_B$.

It is hard to measure the triple differential decay and several integrations must be carried out. We define the limits of the integrations. For fixed q^2 we can substitute q_0 by the hadronic mass $s = m_X^2$:

$$\begin{aligned} s &= m_B^2 + q^2 - 2m_B q_0, \\ &= m_B^2 + q^2 - 2m_B E_e - 2m_B E_\nu. \end{aligned} \tag{14.15}$$

We can replace E_ν by $E_\nu = q^2/[2E_e(1 - \cos\theta)]$ and obtain the upper bound of s at $\cos\theta = -1$:

$$m_\pi^2 \le s \le (m_B - 2E_e)\left(m_B - \frac{q^2}{2E_e}\right). \tag{14.16}$$

The other two variables are bounded in the regions

$$0 \le q^2 \le 2m_B E_e, \tag{14.17}$$

$$0 \le E_e \le \frac{m_B}{2}. \tag{14.18}$$

The triple differential rate can be used to extract the distribution function directly, but decay rates in several variables are not available yet. Instead one adopts an *Ansatz* for the distribution function. Here we are guided by physical intuition and the experience gained from the fragmentation functions. In the boosted frame, we expect the heavy quark to carry most of the momentum of the meson, which means that the distribution function is peaked at $x \approx 1$. Thus a function with a peak in the high-x region and a small width should be sufficient. A function with two parameters a and b that satisfies the above criteria is

$$f(x) = N\frac{x(1-x)}{(x-b)^2 + a^2}, \tag{14.19}$$

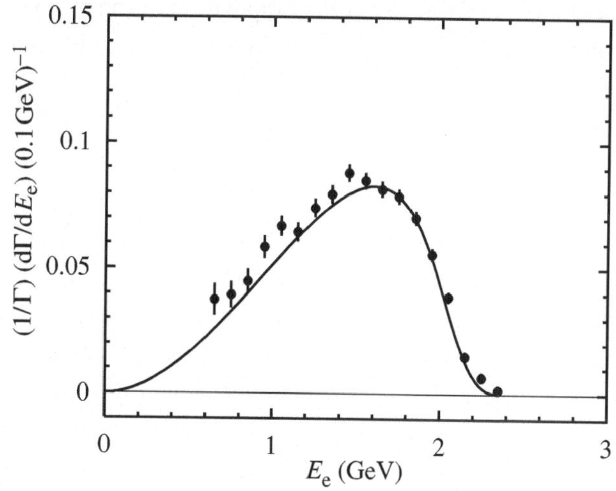

Figure 14.3. The predicted electron-energy spectrum compared with data.

with N a normalization constant. Integration over the variables and for $a = 0.0118$, $b = 0.931$ produces the electron spectrum shown in Fig. 14.3. A similar curve is obtained in the ACCMM model. The data points are from Barish *et al.* (1996).

It is desirable to calculate the distribution function or determine its parameters from basic principles. Attempts in this direction have been made in the heavy-quark effective theory.

14.3 Exclusive semileptonic decays

Of special interest are decays in which the initial and final mesons are specific hadrons. In this case analysis in the quark-decay picture is not applicable. Let us concentrate on decays of a heavy B or D meson, to a meson X_i and a lepton pair:

$$\mathrm{M} \to \mathrm{X}_i \mathrm{e}\nu, \tag{14.20}$$

where M is the initial heavy meson. The experiments measure the energy spectrum of the electron $(1/\Gamma)\mathrm{d}\Gamma/\mathrm{d}E$, and, for comparison with the theoretical model, we calculate the invariant amplitude for the process. It is again a product of the known leptonic weak current L^μ and the hadronic matrix element H^μ:

$$\mathcal{M} = \frac{G}{\sqrt{2}} L^\mu H_\mu \quad \text{with} \quad H^\mu = \langle X_i | j^\mu | M \rangle. \tag{14.21}$$

The explicit structure of the meson current depends on the definite properties of the final meson under Lorentz transformations. With such information it is possible to derive the most general form-factor decomposition of the matrix element. As an example we present here the decay to a pseudoscalar state, i.e. $X_i = P(0^-)$:

$$\langle P|j_\mu|M\rangle = \frac{m_{\mathrm{M}}^2 - m_{\mathrm{P}}^2}{q^2} q_\mu F_0(q^2) + \left[(k_{\mathrm{M}} + k_{\mathrm{P}})_\mu - \frac{m_{\mathrm{M}}^2 - m_{\mathrm{P}}^2}{q^2} q_\mu\right] F_1(q^2), \tag{14.22}$$

with $q_\mu = (k_{\mathrm{M}} - k_{\mathrm{P}})_\mu$ and $F_0(0) = F_1(0)$. $F_0(q^2)$ and $F_1(q^2)$ denote the longitudinal and transverse form factors, respectively. The form factors are unknown parameters, which have to be estimated theoretically and then compared with the experiments. On the theoretical side, the most famous parametrization appears in the BSW model (Wirbel *et al.*, 1985) which makes the following *Ansatz* for the q^2 dependence:

$$F_i(q^2) = \frac{F_i(0)}{1 - q^2/m_i^2}. \tag{14.23}$$

The pole masses m_i are estimated numerically. Finally, to predict an energy spectrum for the lepton produced in the decay of Eq. (14.20), an estimate for the form factors at $q^2 = 0$ is required. In the BSW appoach it is given by the overlap of the initial and final wave functions of the mesons. With this information, the phase-space integrals can be performed and many differential decay rates calculated (Wirbel *et al.*, 1985).

14.4 Heavy-quark effective theory

The heavy-quark effective theory (HQET) is a systematic method for describing particles containing a heavy quark (Manohar and Wise, 2000). The HQET is based on the fact that QCD is flavor-blind and a hadron containing a heavy quark is very unlikely to have excitations containing antiquark degrees of freedom. Thus it suffices to work with the spinor of the heavy quark and view the light quark with its surrounding gluonic field as a composite system. This picture is an improvement over the spectator model and supplies a method for calculating the distribution function of the heavy quark in the meson. For mesons made up of a heavy quark Q and a light antiquark, the heavy quark is essentially on-shell and therefore static when the meson is at rest. This is analogous to atoms, in which the nucleus is stationary and the electrons move in the static field of the nucleus. We will see that, to zeroth order, properties of the meson do not depend on the mass, spin, or flavor of the heavy quark, just as chemical properties of the atom do not depend on the particular isotope.

We shall demonstrate that the HQET is a limiting case of QCD where $m_{\mathrm{Q}} \to \infty$ with the four-velocity of Q, v_μ, held fixed. In this limit the Lagrangian simplifies considerably, giving a static term and calculable corrections expressed in inverse powers of the heavy-quark mass.

The four-momentum of the heavy quark can be expressed by

$$p_\mu = m_Q v_\mu + k_\mu, \tag{14.24}$$

with v_μ its velocity normalized to unity, $v_\mu v^\mu = 1$. The correction term k_μ is small relative to $m_Q v_\mu$. The heavy-quark field has an energy dependence close to that of a free particle and a spinor h_v with positive energy. Since the heavy quark is bound, it may be influenced by the cloud of quark–antiquark pairs. These degrees of freedom are represented in HQET by two spinors, which are eigenfunctions of the velocity operator. To be specific,

$$Q(x) = \mathrm{e}^{-\mathrm{i}m_Q v\cdot x}[h_v(x) + l_v(x)]\,, \tag{14.25}$$

with

$$\not{v} h_v = h_v, \qquad \not{v} l_v = -l_v. \tag{14.26}$$

Factoring out the phase is a redefinition of the field, but the decomposition into the components h_v and l_v is an approximation in terms of positive and negative velocity fields.

The results can be re-expressed in terms of velocity projection operators

$$P_\pm = \frac{1}{2}(1 \pm \not{v}), \tag{14.27}$$

which project out the positive velocity fields,

$$P_+ Q = \mathrm{e}^{-\mathrm{i}m_Q v\cdot x} h_v, \tag{14.28}$$

and the negative velocity fields,

$$P_- Q = \mathrm{e}^{-\mathrm{i}m_Q v\cdot x} l_v. \tag{14.29}$$

The overall phase varies rapidly and among the spinors we expect h_v to be dominant.

The heavy quark interacts with the gluonic field and satisfies a Dirac Lagrangian

$$\mathcal{L} = \bar{Q}(x)\big[\mathrm{i}\not{D} - m_Q\big]Q(x), \tag{14.30}$$

where $D_\mu = \partial_\mu + \mathrm{i}g_s A^a_\mu \cdot \lambda^a/2$, with $A^a_\mu(x)$ the field of the gluons. We are interested in approximate solutions of the Lagrangian in which the heavy-quark field h_v is slightly perturbed by the light degrees of freedom. To simplify the algebra we mention two identities and drop the subscript v, i.e. setting $h_v = h$ and $l_v = l$,

$$\begin{aligned} P_+\gamma_\mu P_+ &= P_+\big(P_-\gamma_\mu + v_\mu\big) = v_\mu P_+, \\ P_-\gamma_\mu P_- &= P_-\big(P_+\gamma_\mu - v_\mu\big) = -v_\mu P_-, \end{aligned} \tag{14.31}$$

$$\not{D}\, Q(x) = -\mathrm{i}m_Q \not{v} Q + \mathrm{e}^{-\mathrm{i}m_Q v\cdot x}(\not{D}h + \not{D}l)\,, \tag{14.32}$$

with $\bar{h}l = 0$ and $\bar{h}\not{v}l = \bar{l}\not{v}h = 0$. With the help of the identities, we rewrite $\mathcal{L}$ in the following way:

$$\begin{aligned}\mathcal{L} &= \bar{h}\mathrm{i}\not{D}\,h + \bar{l}(\mathrm{i}\not{D} - 2m_{\mathrm{Q}})l + \bar{h}\mathrm{i}\not{D}l + \bar{l}\mathrm{i}\not{D}h \\ &= \mathrm{i}\bar{h}v\cdot Dh - \bar{l}(\mathrm{i}v\cdot D + 2m_{\mathrm{Q}})l + \bar{h}\mathrm{i}\not{D}l + \bar{l}\mathrm{i}\not{D}h.\end{aligned} \tag{14.33}$$

In the last two terms a final simplification is possible, which allows one to replace D_μ by its transverse component $D_{\perp\mu} = D_\mu - D\cdot vv_\mu$ because $\bar{h}\not{v}l = \bar{l}\not{v}h = 0$. Several authors use this additional simplification. The Euler–Lagrange equation for the field gives

$$(\mathrm{i}v\cdot D + 2m_{\mathrm{Q}})l = \mathrm{i}\not{D}h \Longrightarrow l = \frac{1}{\mathrm{i}v\cdot D + 2m_{\mathrm{Q}}}\,\mathrm{i}\not{D}h. \tag{14.34}$$

To sum up, we have taken out the dependence on large momenta by factoring out the term $\exp(-\mathrm{i}m_{\mathrm{Q}}v\cdot x)$. The remaining terms, except for the covariant derivative acting on h, contain small momenta, as is evident from the fact that $l(x)$ is indeed the smaller of the two velocity spinors.

In this limit of the theory, we give rules for the propagator

$$\mathrm{i}\frac{\not{p} + m_{\mathrm{Q}}}{p^2 - m_{\mathrm{Q}}^2} = \mathrm{i}\frac{m_{\mathrm{Q}}\not{v} + m_{\mathrm{Q}} + \not{k}}{2m_{\mathrm{Q}}v\cdot k + k^2} \to \frac{\mathrm{i}}{v\cdot k}\left(\frac{1+\not{v}}{2}\right) + \mathcal{O}\left(\frac{k}{m_{\mathrm{Q}}}\right) \tag{14.35}$$

and for the vertex describing the coupling of gluons to the heavy quark,

$$\mathrm{i}g\bar{h}\gamma_\mu h = \mathrm{i}gv_\mu\bar{h}h. \tag{14.36}$$

Calculations with h alone reproduce the free-quark model. The interactions are improved by including $1/m_{\mathrm{Q}}$ terms.

In the limit of $m_{\mathrm{Q}} \to \infty$ the quark field approaches the heavy-quark field,

$$Q(x) \to \mathrm{e}^{-\mathrm{i}m_{\mathrm{Q}}v\cdot x}P_+h. \tag{14.37}$$

On substituting this field into the Lagrangian (14.30) or from Eq. (14.33), we obtain

$$\mathcal{L} \to \mathrm{i}\bar{h}\,v\cdot Dh\; = \mathcal{L}_0. \tag{14.38}$$

The result shows explicitly that in the limit $m_{\mathrm{Q}} \to \infty$ the interaction between the gluonic field and the heavy quark is independent of the spin and mass of the heavy quark. There are several consequences of this property.

As mentioned already, the meson is viewed as composed of a heavy quark and a light system consisting of the light quark and its surrounding gluonic field. Several properties should be independent of the spin of the heavy quark. Let us denote the spin of the light system by S_ℓ and the spin of the heavy quark by $S_Q = \frac{1}{2}$. The two systems may still have relative angular momentum $\vec{L}$. The total angular momentum, $\vec{J}$, for the meson is a sum of spins

$$\vec{S} = \vec{S}_Q + \vec{S}_\ell \tag{14.39}$$

plus the relative angular momentum $\vec{L}$,

$$\vec{J} = \vec{S} + \vec{L}.$$

For a meson with $S_\ell = \frac{1}{2}$ and $\vec{L} = 0$

$$|\vec{S}| = |\vec{S}_\ell| \pm \frac{1}{2} = 0,\ 1$$

and the total angular momentum $\vec{J} = 0$ or 1. For Q = b the two states with $L = 0$ are B_d and B_d^* mesons. The HQET predicts that the two states have properties that are independent of S_Q and therefore degenerate. The difference between the masses of the pseudoscalar and vector mesons is indeed small relative to their sum. The same is experimentally true for the charm states D and D^*.

Even though the mass differences are small, they are not zero, suggesting that we must formulate the effective Lagrangian more precisely by keeping the $1/m_Q$ terms which were ignored. This will be discussed in the next section. The same is true for the charm states D and D^* containing charm quarks. To first approximation the spin symmetry is realized. However, the small differences suggest that we must formulate the effective Lagrangian more accurately by including $1/m_Q$ corrections.

14.5 The effective Lagrangian: $1/m_Q$ corrections

Higher-order corrections are obtained by including the light-quark degrees of freedom. This complicates the algebra somewhat, but leads to correction terms that are simple to describe. In the following we give a derivation. However, the reader who finds it complicated may proceed directly to Eq. (14.43).

To include the light-quark field, we substitute $l(x)$ using Eq. (14.34) together with a similar expression for $\bar{l}(x)$. In this way we obtain an improved interaction,

$$\begin{aligned}\mathcal{L} = \mathcal{L}_0 + \mathcal{L}_1 &= \bar{h}\left(\mathrm{i}\not{D} + \mathrm{i}\not{D}\frac{1}{\mathrm{i}v \cdot D + 2m_Q}\mathrm{i}\not{D}\right)h + \mathcal{O}\left(\frac{1}{m_Q^2}\right)\\ &= \mathrm{i}\bar{h}v \cdot Dh + \frac{1}{2m_Q}\bar{h}\mathrm{i}\not{D}\mathrm{i}\not{D}h + \mathcal{O}\left(\frac{1}{m_Q^2}\right),\end{aligned} \tag{14.40}$$

with the first term being $\mathcal{L}_0$ and the rest the improvement. The new term has two covariant derivatives next to each other and can be simplified. We use the identity

$$\begin{aligned} \not{D}\not{D} &= \gamma^\mu \gamma^\nu D_\mu D_\nu \\ &= \frac{1}{2}\{\gamma^\mu, \gamma^\nu\} D_\mu D_\nu + \frac{1}{2}[\gamma^\mu, \gamma^\nu] D_\mu D_\nu \\ &= D^2 + \frac{1}{4}[\gamma^\mu, \gamma^\nu][D_\mu, D_\nu] \end{aligned} \tag{14.41}$$

together with

$$\sigma^{\mu\nu} = \frac{\mathrm{i}}{2}[\gamma^\mu, \gamma^\nu], \qquad \mathrm{i}g_s G^a_{\mu\nu}\left(\frac{\lambda_a}{2}\right) = [D_\mu, D_\nu], \tag{14.42}$$

in order to obtain the $1/m_Q$ term of the Lagrangian:

$$\mathcal{L}_1 = -\bar{h}\frac{D^2}{2m_Q}h - g_s\bar{h}\frac{\sigma^{\mu\nu}G_{\mu\nu}}{4m_Q}h. \tag{14.43}$$

A few remarks are now in order. The first term, $D^2/(2m_Q)$, is the average kinetic energy of the heavy quark. The second term couples the spin of the heavy quark to the gluo-magnetic field surrounding the quarks. The theory succeeded in including the interaction of the heavy quark with the surrounding degrees of freedom. It also has the advantage that it is precise enough to allow systematic studies. It has been successful in accounting for mass differences, decay rates etc., which we cover in the next section.

14.5.1 Applications

Mass relations

To leading order the HQET Lagrangian does not depend on the spin or flavor of the heavy quark. The first term in Eq. (14.43) breaks the flavor symmetry and the second term breaks both flavor and spin symmetries. These two terms play a leading role in determining the mass spectra of heavy hadrons. Consider the expectation value of the Hamiltonian between mesonic states. The mass of the meson is written as

$$\begin{aligned} m_M &= m_Q + \bar{\Lambda} + \langle M|O_1|M\rangle + \langle M|O_2|M\rangle \\ &= m_Q + \bar{\Lambda} - \frac{\lambda_1}{2m_Q} + \frac{a_J\lambda_2}{2m_Q}, \end{aligned} \tag{14.44}$$

with $a_J = \frac{1}{2}\vec{S}_Q \cdot \vec{S}_\ell$. Here O_1 and O_2 are the first and second operators in Eq. (14.43), respectively, and $\bar{\Lambda}$ is the contribution from the light-quark degrees of freedom, which is a priori unknown. According to HQET, λ_1 and λ_2 are universal constants. One generally determines them from the moments of the energy spectrum of inclusive semileptonic B decays. It turns out that $\bar{\Lambda} \approx 0.40$ GeV and $\lambda_1 \approx -0.2$ GeV^2.

O_2 is a magnetic-moment-type interaction, and its matrix element for mesons with $L = 0$ is proportional to $S_Q \cdot S_\ell = (J^2 - S_Q^2 - S_\ell^2)/2$. The expectation values of this operator in triplet and singlet states are different, which causes a splitting among the lowest-lying states. The singlet ($J = 0$) is lowered and the triplet ($J = 1$) is raised, but by different amounts. These states are generally denoted by M and M^*. It is obvious that

$$\frac{m_{B^*} - m_B}{m_{D^*} - m_D} = \frac{m_c}{m_b},$$

which is satisfied to a high degree of accuracy. In fact, one can determine λ_2 from the splitting of the meson doublet.

Exclusive semileptonic decays

An important prediction of HQET describes decays governed by the quark-level transition $b \to c$. The initial B meson decays into an electron–neutrino pair plus the D meson. In general the matrix element of pseudoscalar-to-pseudoscalar transition involves two form factors that are functions of q^2 (see Eq. (14.22)). The velocities for the B and D mesons are $v^\mu = p^\mu/m_B$ and $v'^\mu = p'^\mu/m_D$ and the q^2 dependence may be replaced by

$$w = v \cdot v' = \frac{m_B^2 + m_D^2 - q^2}{2m_B m_D}. \tag{14.45}$$

In the limit that the masses are very big relative to the difference $m_B - m_D$ the two states B and D are, according to HQET, very similar. In this limit $v \cdot v' = 1$ and $v^\mu \approx v'^\mu$.

To appreciate the situation, consider a configuration in which the B meson is at rest and the four-momentum carried away by the lepton pair is maximal, $q^2 = (M_B - M_D)^2$, which means that the D meson is also produced at rest. This is the zero-recoil point where $w = 1$. In the limit that the heavy-meson masses are large and close to each other, only one form factor contributes:

$$\begin{aligned} \langle D(p')|V_\mu|B(p)\rangle &= f_+(q^2)(p + p')_\mu + f_-(q^2)(p - p')_\mu \\ \to \langle D(v')|\bar{c}\gamma_\mu b|B(v)\rangle &\approx \sqrt{m_B m_D}\, h(v \cdot v')(v_\mu + v'_\mu). \end{aligned} \tag{14.46}$$

The square root arises from the normalization of the states and $h(v \cdot v')$ is a universal function called the Isgur–Wise function (Isgur and Wise, 1989, 1990). Furthermore, in the limit $m_B, m_D \to \infty$ with $m_B - m_D$ fixed, $q^2 \ll M_B, M_D$, the Isgur–Wise function measures the overlap of two identical wave functions so that $h(1) = 1$.

Consider next the decay to a vector meson $B \to D^*\ell\nu$. We have discussed already the fact that to leading order the decay is independent of the spin of the heavy quarks. A consequence of the symmetry is the relation

$$\langle D^*(v', \varepsilon)|\bar{c}\gamma_\mu b|B(v)\rangle = \mathrm{i}\sqrt{m_B m_{D^*}}\epsilon_{\mu\nu\alpha\beta}\epsilon^\nu v^\alpha v'^\beta h(v \cdot v'), \tag{14.47}$$

where the Lorentz structure depends on the polarization of the D^* meson, ϵ^μ, and the same universal form factor $h(v \cdot v')$ appears as before. However, the universality is a feature of the zeroth-order Lagrangian and is broken by $1/m_Q$ corrections. For a complete analysis we must also include matrix elements of the axial current, which bring additional form factors into play. The general case must explain decays to several vector mesons $D^*(2010)$, $D^*(2420)$, ..., whose partial decay widths are different from each other. The differences remain after corrections over phase-space factors have been taken into account.

This motivated an analysis of the partial decay rates in terms of $h(v \cdot v')$ plus its slope at the point $v \cdot v' = 1$. The important variable for the decays is w, with the decay spectrum given by (Manohar and Wise, 2000)

$$\frac{\mathrm{d}\Gamma}{\mathrm{d}w}(B \to D^*\ell\nu) = \frac{G^2|V_{cb}|^2}{48\pi^3}K(w)h(v \cdot v')^2, \tag{14.48}$$

where w is defined in (14.45), $K(w)$ is a known phase-space factor, and $h(v \cdot v')$ is the Isgur–Wise function including finite-mass and other (QCD, ...) corrections. The analysis determined the value of the form factor $F_1(w = 1)$ and its slope at $w = 1$. A systematic analysis of several experiments gives the mean value of the slope and the product as

$$h(w = 1)|V_{cb}| = 0.0038 \pm 0.0010,$$

from which the value for $|V_{cb}|$ has been extracted (see Section 9.3.4).

The applications of HQET to mass relations and B decays to D and D^* are successes of the theory. The Isgur–Wise function $h(v \cdot v')$ has been studied and calculated also in lattice gauge theories. Of direct interest is its calculation for the physical b and c masses in the zero-recoil limit (when $v \cdot v' = 1$). Lattice calculations give $h(1) = 0.929$ with a small error.

B-meson decays to light mesons are much harder to calculate in HQET or on the lattice because most of the decay occurs in a kinematic region where the π or the ρ mesons have large momenta. They attract a good deal of attention because data on B-meson decays are accumulating in B-factories and are very important for the interpretation of CP asymmetries in B-meson decays.

Inclusive semileptonic decays

For inclusive decays one must calculate the decay rate in the rest frame of the b quark. To leading order the hadronic tensor will have the structure of Eq. (14.10), which is proportional to the δ-function

$$\delta[(p_{\mathrm{b}} - q)^2] = \delta\left(m_{\mathrm{b}}^2 - 2p_{\mathrm{b}} \cdot q + q^2\right).$$

In the limit $m_{\mathrm{b}}^2 \gg q^2$ it reduces to

$$\delta\left(m_{\mathrm{b}}^2 - 2m_{\mathrm{b}} q_0\right),$$

which peaks at the endpoint of the q_0 range. Since there is no averaging over a distribution function, the smoothness of the spectrum must be brought about by higher-order corrections of HQET. Higher-order corrections including bound-state effects must reproduce the effects and spectrum discussed in Sections 14.2.1 and 14.2.2.

14.6 The top quark and its physical properties

The top quark is very heavy and weighs as much as a heavy atom. After the discovery of the bottom quark and the tau lepton, the top was predicted in order to preserve the symmetry between quarks and leptons. In addition, several properties of the low-lying states required the existence of a heavier state. Among these properties are the mixing and CP properties of the neutral K mesons, as well as those of neutral D and B mesons. Furthermore, in order to accommodate CP violation in the electroweak theory, a third generation of quarks is required. Some of these properties were mentioned earlier and they will be covered in greater detail in the next chapters.

Besides the expectation of a heavier top quark, its mass was, for a long time, unknown. Only with the discovery of the W and Z bosons and the precise measurements of their masses and widths did it become possible to put a limit on the top quark's mass. A stringent limit is provided by the ρ parameter, where the correction from the self-energy has a quadratic dependence on the quark mass. We quote the result here in order to stress how quantum corrections become important. The ρ parameter is defined as

$$\rho = \frac{M_{\mathrm{W}}^2}{M_{\mathrm{Z}}^2 \cos^2\theta_{\mathrm{W}}}$$

and has the value unity (see Eq. (12.19)). The loop correction from the top–bottom quark pair changes this ratio to

$$\rho = 1 + \frac{G}{8\pi^2}\left(m_{\mathrm{t}}^2 + m_{\mathrm{b}}^2 - \frac{2m_{\mathrm{t}}^2 m_{\mathrm{b}}^2}{m_{\mathrm{t}}^2 - m_{\mathrm{b}}^2}\ \ln\left(\frac{m_{\mathrm{t}}^2}{m_{\mathrm{b}}^2}\right) + \begin{matrix}\text{smaller terms from}\\ \text{Higgs exchanges}\end{matrix}\right). \tag{14.49}$$

For very accurate values of M_{W}, M_{Z}, and the mixing angle, the effect from the top quark is noticeable. Several analyses along these lines gave a range for the top quark's mass in the neighborhood of 175 GeV, with an error of ±25 GeV, within which it was eventually discovered.

The dominant decay of the top quark is into a bottom quark plus a W boson. The calculation of the decay is straightforward and we have formulated it as an exercise. If we neglect the mass of the bottom quark, the decay width is

$$\Gamma(\mathrm{t} \to \mathrm{bW}^+) = \frac{G_{\mathrm{F}} M_{\mathrm{t}}^3}{8\sqrt{2}\pi} |V_{\mathrm{tb}}|^2 \left(1 - \frac{M_{\mathrm{W}}^4}{M_{\mathrm{t}}^4}\right)^2, \tag{14.50}$$

which grows rapidly with the top-quark mass. For $M_{\mathrm{t}} = 175\,\mathrm{GeV}$, the decay width is

$$\Gamma(\mathrm{t} \to \mathrm{bW}^+) = 1.55\,\mathrm{GeV},$$

which corresponds to a top-quark lifetime of 0.4×10^{-24} s. The confining effects of the strong interactions act on a time scale $\approx 1/\Lambda_{\mathrm{QCD}}$. This means that the top quark decays long before the interaction can act to produce hadrons. Unlike the properties of the other five quarks, there are no bound states and no toponium spectroscopy.

The fact that the mass of the top quark is close to the value required by the radiative corrections is a success of the electroweak theory. These and other tests of the theory will be discussed in Section 17.2 devoted to precision tests of the theory. Another test involves the value of V_{tb}, which, according to the analysis of the CKM matrix, must be very close to unity. There is already experimental evidence that the top decays primarily to a bottom quark, but the accuracy is still very poor to test the unitarity prediction in Eq. (9.49).

The ultrarapid decay of the top quark implies that the production of $\mathrm{t\bar{t}}$ pairs in hadron collisions should be calculable in perturbative QCD. Electron–positron colliders could be used to search for the reaction

$$\mathrm{e^+e^-} \to \mathrm{t\bar{t}},$$

but at the lower energies available in those colliders it is impossible to produce the pairs. The decisive experiments have been carried out at the proton–antiproton collider of Fermilab. Some characteristic reactions and their decays are listed below:

$$
\begin{aligned}
\mathrm{p}+\bar{\mathrm{p}} \to \mathrm{t}+\bar{\mathrm{t}}+\text{anything} &\to \mathrm{e}\nu_{\mathrm{e}} + \mu\bar{\nu}_{\mu} + \text{hadrons},\\
&\to \mathrm{e}^{\pm}\mu^{\pm}\,\mathrm{b}\bar{\mathrm{b}}\,\not{E}_{\mathrm{T}},\\
&\to \mu^{\pm} + \text{jets}\,\not{E}_{\mathrm{T}}.
\end{aligned}
$$

The dilepton events (eμ, ee, and $\mu\mu$) are produced when both W bosons decay into eν or $\mu\nu$. The neutrinos remain unobservable and are represented by the missing energy $\not{E}_{\mathrm{T}}$. Events with lepton-plus-jets channels occur when one W decay produces e or μ and the other decays into quark–antiquark pairs.

A second challenge to experimenters is the complexity of the events in high-energy proton–antiproton collisions. The $\mathrm{t}\bar{\mathrm{t}}$ pair produced is accompanied by scores of other particles. Separating the top quark is like searching for a needle in a haystack. Two experimental groups at the Tevatron collider at Fermilab succeeded in discovering the production of $\mathrm{t}\bar{\mathrm{t}}$ pairs. As mentioned in the introduction to this chapter, the production of $\mathrm{t}\bar{\mathrm{t}}$ pairs and their subsequent decays are computed in the parton model, using the best possible quark distribution functions, and the top quark's mass is determined as the value which optimizes the fit.

We close this section with a speculation. The fact that the mass of the top quark is larger than the mass of gauge bosons and closer to the scale of the symmetry-breaking motivates the thought that the top quark is intimately connected to the symmetry-breaking. One suggestion is to study the decays of the top quark to the b quark and the W boson, which is expected to be longitudinally polarized. Since the longitudinal state of W bosons is developed through the Higgs mechanism, it may be sensitive to new physics. In another suggestion, the top–antitop pairs attract each other through a new force to make a condensate, which is the Higgs particle. We return to this possibility in Chapter 17.

14.7 Loop diagrams with heavy quarks

14.7.1 Mixing of states and lifetime differences: a preview

Heavy quarks also appear as intermediate states in loop diagrams and give dominant contributions. As a first example we study box diagrams, in which second-order weak interactions change the flavor quantum number by two units and produce the mixing of B^0–$\bar{\mathrm{B}}^0$, as well as K^0–$\bar{\mathrm{K}}^0$, states. When the intermediate states are heavy quarks, they produce a short-distance interaction shown with the diagrams in Fig. 14.4. The strategy of the calculations is to construct an effective $\Delta F = 2$ Lagrangian from the free-quark model and then take the matrix element between B^0 and $\bar{\mathrm{B}}^0$ (or, analogously, K^0 and $\bar{\mathrm{K}}^0$). We compute the diagram of Fig. 14.4 in

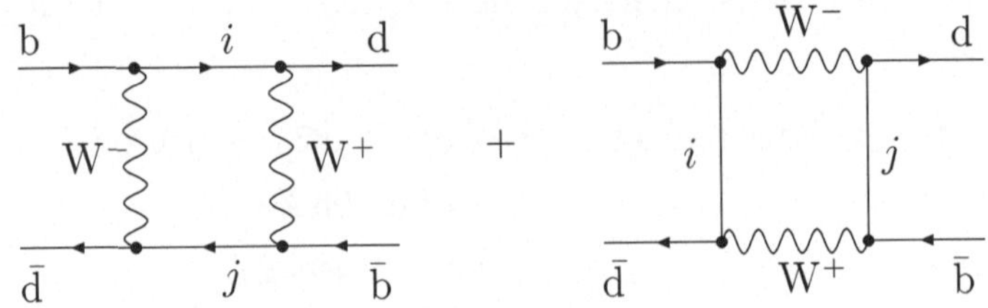

Figure 14.4. Box diagrams for $\Delta B = 2$ transitions. We calculate the "scattering" term in (a). The "annihilation" term (b) is equal to the "scattering" term.

Section 14.7.2, where it is also shown that box diagrams are finite, i.e. they do not bring in any ultraviolet divergences. Before we present the calculation, we quote several results from Chapters 15 and 16 in order to emphasize the point that box diagrams are indeed important. The interested reader may consult Chapters 15 and 16 for the underlying physics and then return to this section for the study of a loop diagram.

Let us denote by $\Delta M_\mathrm{d} = M_{\mathrm{B_H}} - M_{\mathrm{B_L}}$ the difference in mass between the physical states built up from the mesons $\mathrm{B_d^0}$ and $\bar{\mathrm{B}}_\mathrm{d}^0$. We use the subscripts H and L to denote heavy and light mesonic states. The structure of physical states is described in Chapter 15. The calculation of the box diagram with the top quark in the intermediate state gives

$$\Delta M_\mathrm{d} = \frac{G^2}{16\pi^2}|V_\mathrm{td}V_\mathrm{tb}^*|^2 M_\mathrm{W}^2 E(x_\mathrm{t}) X_{B_\mathrm{d}}, \quad \text{with} \quad x_\mathrm{t} = \left(\frac{m_\mathrm{t}}{m_\mathrm{W}}\right)^2, \tag{14.51}$$

$$E(x_\mathrm{t}) = \frac{3}{2}\frac{x_\mathrm{t}^3}{(1-x_\mathrm{t})^3}\ln x_\mathrm{t} - \left[\frac{1}{4} + \frac{9}{4}\frac{1}{1-x_\mathrm{t}} - \frac{3}{2}\frac{1}{(1-x_\mathrm{t})^2}\right]x_\mathrm{t}, \tag{14.52}$$

$$\bar{X}_{\mathrm{B_d}} = \langle B_\mathrm{d}|\bar{d}\gamma_\mu(1-\gamma_5)b\bar{b}\gamma^\mu(1-\gamma_5)d|\bar{B}_\mathrm{d}\rangle. \tag{14.53}$$

The function $E(x_\mathrm{t})$ comes from the integration over the loop and $X_{\mathrm{B_d}}$ is the matrix element of a four-quark operator between $\mathrm{B_d}$ and $\bar{\mathrm{B}}_\mathrm{d}$ states. Four-quark operators appear as overall factors in calculations of box diagrams. The same calculation also determines the width difference

$$\Delta\Gamma_\mathrm{d} = \frac{3G^2}{32\pi} m_\mathrm{b}^2 |V_\mathrm{td}V_\mathrm{tb}^*|^2 X_{\mathrm{B_d}}. \tag{14.54}$$

In the decay only u and c quarks appear as final states. The decay is discussed at the end of Section 16.4 and shown in Fig. 16.1. The dependences of the width and ΔM_d on the quark masses are different. Mass and width differences are related by

$$\Delta\Gamma_\mathrm{d} = \frac{3}{2}\pi\frac{m_\mathrm{b}^2}{M_\mathrm{W}^2 E(x_\mathrm{t})}\Delta M_\mathrm{d}, \tag{14.55}$$

from which we conclude that for large m_t the width difference is much smaller than the mass difference. Furthermore, on comparing ΔM_d with the decay width

$$\Gamma_B = \frac{G^2 m_b^5}{192\pi^3} |V_{bc}|^2, \tag{14.56}$$

we note that there is an enhancement of the mass difference arising from the high mass of the heavy quark. Thus, for favorable values of the other quantities, ΔM_d can be comparable to Γ_B. In fact, experiments determined

$$\Delta M_d = (0.49 \pm 0.01) \times 10^{12} \hbar\, s^{-1}, \tag{14.57}$$

which is one fifth of the width. A similar analysis of the mass difference of $B_s = (b\bar{s})$ leads to a larger mass difference because it is enhanced by the CKM element. Experimentally there is the lower bound

$$\Delta M_s > 13.1 \times 10^{12} \hbar\, s^{-1}. \tag{14.58}$$

The discussion shows that loop contributions are important since they determine physical quantities and correlate properties of various mesons.

The physics described for the neutral B_d mesons contains several general properties. The down quarks are always lighter than the upper quark of the same family. Consequently, mesons containing down quarks decay to quarks of a lighter family, for which the decay width is suppressed by a CKM element. On the other hand, the mass difference of neutral mesons containing the down quarks involves the square of the mass of the upper quark. If the mass of the upper quark in the same family is much larger and the values of the mixing angles and the reduced matrix element are favorable, then substantial mixing between the neutral mesons is possible. This situation is realized for the B_d^0, B_s^0, and K_L^0–K_S^0 mesons.

The opposite situation prevails for neutral mesons that contain a heavy upper quark. For example, for $D^0 = (c\bar{u})$ mesons the decay width is proportional to m_c^5 and the CKM element is practically unity. The mass difference is proportional to m_b^2 or m_s^2 and the CKM elements are smaller than unity, so there is no substantial enhancement. Consequently, mixing of the D^0 and $\bar{D}^0$ states is small and it has not been observed yet.

14.7.2 Calculation of a box diagram

We compute first the box diagrams for Fig. 14.4 with intermediate quarks $i, j = u, c, t$. In addition to the diagrams with the exchange of W bosons there are exchanges with charged Higgses, which become important when the masses of the internal quarks are large. Charged Higgses became the longitudinal degrees of freedom of the W bosons. For simplicity we shall assume that the external momenta and masses are zero. In the Feynman–'t Hooft gauge, we obtain

$$T_{12}^{a} = \left(\frac{g}{2\sqrt{2}}\right)^4 \sum_{i,j} \lambda_i \lambda_j \int \frac{\mathrm{d}^4 k}{(2\pi)^4} \left(\frac{-\mathrm{i}}{k^2 - M_{\mathrm{W}}^2}\right)^2 \left(\bar{d}_{\mathrm{L}} \gamma^{\mu} \frac{\not{k} + m_i}{k^2 - m_i^2} \gamma^{\nu} b_{\mathrm{L}}\right)$$
$$\times \left(\bar{d}_{\mathrm{L}} \gamma_{\nu} \frac{\not{k} + m_j}{k^2 - m_j^2} \gamma_{\mu} b_{\mathrm{L}}\right)$$
$$= -\frac{g^4}{64} \sum_{i,j} \lambda_i \lambda_j \int \frac{\mathrm{d}^4 k}{(2\pi)^4} \frac{1}{\left(k^2 - M_{\mathrm{W}}^2\right)^2} \bar{d} \gamma^{\mu} (1 - \gamma_5) \frac{\not{k} + m_i}{k^2 - m_i^2} \gamma^{\nu} (1 - \gamma_5) b$$
$$\times \bar{d} \gamma_{\nu} (1 - \gamma_5) \frac{\not{k} + m_j}{k^2 - m_j^2} \gamma_{\mu} (1 - \gamma_5) b, \qquad (14.59)$$

with $\lambda_i = V_{id}^* V_{is}$. Simple counting of the momenta shows that the integral is convergent. The neutrino masses in the numerators do not contribute, because of the $(1 - \gamma_5)$ structure. The surviving integral has the following Lorentz structure:

$$I_{\alpha\beta}(i, j) = \int \mathrm{d}^4 k \frac{k_{\alpha} k_{\beta}}{\left(k^2 - M_{\mathrm{W}}^2\right)^2 \left(k^2 - m_i^2\right)\left(k^2 - m_j^2\right)} = I(m_i, m_j, M_{\mathrm{W}}) g_{\alpha\beta}. \qquad (14.60)$$

The fact that the integral is proportional to the $g_{\alpha\beta}$ simplifies the spinor structure, which takes the form

$$\bar{d} \gamma^{\mu} \gamma^{\alpha} \gamma^{\nu} (1 - \gamma_5) b \cdot \bar{d} \gamma_{\nu} \gamma_{\alpha} \gamma_{\mu} (1 - \gamma_5) b = 4 \bar{d} \gamma^{\alpha} (1 - \gamma_5) b \cdot \bar{d} \gamma_{\alpha} (1 - \gamma_5) b. \qquad (14.61)$$

This and other spinor identities follow from the relation

$$\gamma^{\mu} \gamma^{\lambda} \gamma^{\nu} = g^{\mu\lambda} \gamma^{\nu} + g^{\lambda\nu} \gamma^{\mu} - g^{\mu\nu} \gamma^{\lambda} - \mathrm{i} \varepsilon^{\mu\lambda\nu\rho} \gamma_5 \gamma_{\rho}, \qquad (14.62)$$

as demonstrated in Appendix C.

14.7.3 Integrals in Euclidean space

In the calculation of the integrals we consider the momenta to be Euclidean, by setting

$$k_0 = \mathrm{i} k_4. \qquad (14.63)$$

Then the propagators do not have poles. The four-dimensional volume element becomes

$$\mathrm{d}^4 k = \mathrm{d}k_0 \, \mathrm{d}^3 k = \mathrm{i} \mathrm{d}k_4 \, \mathrm{d}^3 k. \qquad (14.64)$$

A typical integral in loop calculations takes the form

$$I^{\alpha}(m) = \int \mathrm{d}^4k \, \frac{1}{(k^2 - m^2 + \mathrm{i}\varepsilon)^{\alpha}} = \mathrm{i}(-1)^{\alpha} \int_{\text{Eucl.}} \mathrm{d}^4k \, \frac{1}{(k^2 + m^2 - \mathrm{i}\varepsilon)^{\alpha}}. \quad (14.65)$$

In Euclidean space we can transform to spherical coordinates,

$$\mathrm{d}^4k \to k^{4-1} \, \mathrm{d}k \, \mathrm{d}\Omega_4 \quad \text{with} \quad \int \mathrm{d}\Omega_4 = \frac{2\pi^2}{\Gamma(2)} = 2\pi^2. \quad (14.66)$$

For our specific integral

$$\begin{aligned} I(m_i, m_j, M_{\mathrm{W}}) &= \frac{1}{4} \int \mathrm{d}^4k \, \frac{k^2}{\left(k^2 - M_{\mathrm{W}}^2\right)^2 \left(k^2 - m_i^2\right)\left(k^2 - m_j^2\right)} \\ &= \frac{1}{4} 2\pi^2 \int_0^{\infty} \mathrm{d}k \, \frac{k^5}{\left(k^2 + M_{\mathrm{W}}^2\right)^2 \left(k^2 + m_i^2\right)\left(k^2 + m_j^2\right)}. \end{aligned} \quad (14.67)$$

14.7.4 Feynman parameters

The standard way of carrying out momentum integrations in loop integrals is to use the so-called Feynman-parameter technique in order to transform a product of propagators depending on the integration momentum into a single factor. This is accomplished with the following identity:

$$\begin{aligned} \frac{1}{(a_1 + \mathrm{i}\varepsilon)(a_2 + \mathrm{i}\varepsilon)\ldots(a_n + \mathrm{i}\varepsilon)} &= \int_0^1 (n-1)! \, \mathrm{d}x_1 \, \mathrm{d}x_2 \ldots \mathrm{d}x_n \\ &\quad \times \frac{\delta(1 - x_1 - x_2 - \cdots - x_n)}{[a_1x_1 + a_2x_2 + \cdots + a_nx_n + \mathrm{i}\varepsilon]^n}. \end{aligned} \quad (14.68)$$

Simple examples can be worked out. When the propagator $(a_i + \mathrm{i}\varepsilon)^k$, with k an integer, appears on the left-hand side, the corresponding formula is obtained by differentiating the above expression with respect to a_i several times. Following this rule we obtain

$$\frac{1}{(a_1 + \mathrm{i}\varepsilon)(a_2 + \mathrm{i}\varepsilon)(a_3 + \mathrm{i}\varepsilon)} = 6 \int \frac{\mathrm{d}x \, \mathrm{d}y(1 - x - y)}{[a_1(1 - x - y) + a_2x + a_3y + \mathrm{i}\varepsilon]^4}. \quad (14.69)$$

Returning to our integral in Eq. (14.67),

$$\begin{aligned} I(m_i, m_j, M_{\mathrm{W}}) &= \frac{\pi^2}{2} \int_{k=0}^{\infty} \mathrm{d}k \, k^5 \int_0^1 \mathrm{d}x \int_0^{1-x} \mathrm{d}y \\ &\quad \times \frac{1 - x - y}{\left[k^2 + M_{\mathrm{W}}^2(1 - x - y) + m_i^2x + m_j^2y\right]^4}. \end{aligned} \quad (14.70)$$

Now, since the Feynman parameters in the δ-function sum up to unity, the coefficient of the k^2 term in the denominator is always unity. We can perform the $\mathrm{d}k^2$ integration to obtain

$$\begin{aligned} I(m_i, m_j, M_\mathrm{W}) &= \frac{\pi^2}{12M_\mathrm{W}^2}\int_0^1 \mathrm{d}x \int_0^{1-x} \mathrm{d}y \frac{1-x-y}{(1-x-y+x_i^2x+x_j^2y)} \\ &= \frac{\pi^2}{12M_\mathrm{W}^2}\frac{1}{2}\frac{1}{x_i-x_j}\left[\frac{1}{1-x_i}-\frac{1}{1-x_j}+\frac{x_i^2}{(1-x_i)^2}\ln x_i \right. \\ &\qquad \left. -\frac{x_j^2}{\left(1-x_j^2\right)^2}\ln x_j\right], \end{aligned} \tag{14.71}$$

with $x_i = m_i^2/M_\mathrm{W}^2$ and $x_j = m_j^2/M_\mathrm{W}^2$. This diagram gives the dominant contribution for cases in which x_i and $x_j \ll 1$. In the other cases with the exchange of a heavy scalar, additional diagrams must be included; the complete answer is given in Chapter 15.

This first example shows that the calculation of loop diagrams involves four steps.

1. A simplification of the spin structure, which reduces the calculation to a few four-dimensional integrals.
2. The reduction of the denominators to a single factor with the help of Feynman parameters.
3. The completion of the four-dimensional integrals. This is carried through in Euclidean space and holds in the Minkowskian region by analytic continuation. The case of the box diagram is relatively simple, because the integral is convergent. In the general case, the integrals are divergent, which demanded the development of special methods for subtracting the infinities, which respect the gauge invariance of the theory.
4. Finally, there is an integration over the Feynman parameters, which requires special attention at the endpoints, where infrared singularities may occur.

Problems for Chapter 14

1. Draw the graph of $\mathrm{d}\Gamma^0/\mathrm{d}x$, where $x = 2E/W$, in the quark rest frame. Multiply by the QCD correction factor

$$G(x) = \frac{1}{x}2\ln(1-x)[2+\ln(1-x)]$$

and show how the distribution changes.

2. Use the identities given in the text to prove Eq. (14.33). Then show that it is possible to replace D_μ by $D_{\perp\mu} = D_\mu - D\cdot v v_\mu$.
3. For the semileptonic decay $\mathrm{B}\to\mathrm{D}e\bar{\nu}$ the dominant form factor $f_+(q^2)$ is related to the Isgur–Wise function as in Eq. (14.46).

(i) Show that $f_+(q^2)$ is related to $h(v \cdot v')$ as follows:

$$f_+(q^2) = h(v \cdot v') \frac{2\sqrt{m_{\mathrm{B}} m_{\mathrm{D}}}}{m_{\mathrm{B}} + m_{\mathrm{D}}}.$$

(ii) Calculate the decay spectrum $\mathrm{d}\Gamma/\mathrm{d}W$ in the heavy-quark limit.
(Hint: you may begin with $\mathrm{d}\Gamma/\mathrm{d}q^2$ and then take the heavy-quark limit.)

4. Use the Feynman rules from Chapter 8 to calculate the decay width for the top quark given in Eq. (14.50).
5. Prove the following useful identities that involve Feynman parameters:

(i) $$\frac{1}{(a_1 + \mathrm{i}\varepsilon)(a_2 + \mathrm{i}\varepsilon)} = \int_0^1 \frac{\mathrm{d}x}{[a_1 x + a_2(1-x) + \mathrm{i}\varepsilon]^2};$$

(ii) $$\frac{1}{(a_1 + \mathrm{i}\varepsilon)^3(a_1 + \mathrm{i}\varepsilon)} = \int_0^1 \frac{3x^2\,\mathrm{d}x}{[ax + b(1-x) + \mathrm{i}\varepsilon]^4}.$$

References

Altarelli, G., Cabibbo, N., Gorbo, G., Maiani, L., and Martinelli, G. (1982), *Nucl. Phys.* **B208**, 365

Bander, M., Silverman, D., and Soni, A. (1980), *Phys. Rev. Lett.* **44**, 7–9 (for another explanation, see M. Lusignoli and A. Pugliese, hep-ph/0210071)

Bareiss, A., and Paschos, E. A. (1989), *Nucl. Phys.* **B327**, 353

Barish, B., Chandra, M., Chan, S. *et al.* (1996), *Phys. Rev. Lett.* **76**, 1570

Combination of CDF and DO Results on Top Quark Mass (CDF/DO), hep-ex/0507006

Glashow, S. L., Iliopoulos, J., and Maiani, L. (1970), *Phys. Rev.* **D2**, 1285

Isgur, N., and Wise, M. (1989), *Phys. Lett.* **B232**, 113
(1990), *Phys. Lett.* **B237**, 527

Manohar, A. (2000), *Eur. Phys. J.* **C15**, 377–380

Manohar, A. V., and Wise, M. B. (2000), *Heavy Quark Physics* (Cambridge, Cambridge University Press)

Narison, S. (2001), *Phys. Lett.* **B520**, 115; hep-ph/0108065

The CDF collaboration, the DO collaboration and the Tevatron Electroweak Working Group

Wirbel, M., Stech, B., and Bauer, M. (1985), *Z. Phys.* **C29**, 637

Select bibliography

Many topics of heavy-quark theory are presented in the book

Manohar, A. V., and Wise, M. (2000), *Heavy Quark Physics* (Cambridge, Cambridge University Press)

For an overview of top-quark physics, see

Quigg, C. (1997), *Phys. Today*, May, p. 20

15

CP violation: K mesons

15.1 Introduction

Most of the symmetries in elementary-particle physics are continuous. A typical example is the symmetry of rotations around an axis, where the angle of rotation can assume any value between zero and 2π. In addition to continuous symmetries, there are also discrete symmetries, for which the possible states assume discrete values classified with the help of a few integers. For instance, snowflakes exhibit the discrete symmetry of rotations under 60° and crystals exhibit various types of discrete symmetries. In elementary-particle physics there are three discrete symmetries of basic importance: parity, charge conjugation and time-reversal.

Parity is the reflection of space coordinates and will be denoted by P. Under parity there are two states – the object and its space reflection. Parity is familiar from quantum mechanics, where the eigenstates of Hamiltonians are classified according to their properties under space reflection. For spherically symmetric potentials the wave functions are proportional to the spherical harmonics $Y_m^\ell(\theta, \varphi)$ whose parity is $(-1)^\ell$. For a long time it was assumed that the fundamental interactions respect P, but a critical review of experimental evidence led two theoreticians, T. D. Lee and C. N. Yang, to suggest that parity may be violated by the weak interactions. One year later, an experiment led by C. S. Wu brought the proof that the P symmetry is indeed violated by weak interactions.

The symmetry of charge conjugation, to be denoted by C, exchanges particles with antiparticles. One can imagine building an antiworld by replacing all particles by antiparticles. In the antiworld the three interactions gravity, the strong force, and electromagnetism are the same, but the weak interactions are different. For example in the antiworld muon-type antineutrinos are right-handed and produce μ^+ which are also right-handed. In comparison neutrinos are left-handed and always produce, in high-energy reactions, left-handed μ^-. In the weak interactions the C symmetry is broken. However, it was assumed, at that time, that the observed processes do respect the combined CP transformation, the one obtained by applying both C and P transformations.

There is a fundamental reason why CP symmetry plays a crucial role. It is intimately linked to the time-reserval transformation (T). This transformation consists of "looking" at an experiment running backward in time. Although, at the macroscopic level, one can distinguish the real sequence of events from the time-reversed one in terms of large-scale phenomena such as entropy or the expansion of the Universe, this is not a priori evident for microscopic interactions, i.e. it is not a priori evident that the amplitudes for reactions and for the time-reversed reactions are equal.

The analysis of CP violation is facilitated by an important theorem known as CPT theorem. It states that any local field theory based on special relativity and quantum mechanics is invariant under the combined action of C, P, and T. A consequence of the theorem is that CP symmetry implies T symmetry, because any CP violation should be compensated by T violation.

Until 1964 the decays and interactions of particles showed that the CP symmetry was conserved; this created the belief that microscopic phenomena also obey the T symmetry. In 1964 CP violation was observed in an experiment dedicated to the study of K^0 and $\bar{K}^0$ mesons. Since then it has become an active topic of research, with CP violation having been observed so far in the K and the B mesons. In this chapter we study the properties of mesons under discrete symmetries, leaving the study of fermions for more specialized articles and books.

15.2 General properties

We describe now the properties that govern the decays of neutral pseudoscalar mesons, such as K^0, D^0, and B^0_d, when the interactions obey the CPT and CP symmetries. The results guide us to properties of these reactions that indicate breakdown of CP and/or CPT symmetries.

For simplicity of presentation we shall consider the K^0 as an example and describe properties of the K^0–$\bar{K}^0$ system; however, the results are general and hold for the other mesons too. We adopt the phase convention

$$P|K^0\rangle = |K^0\rangle, \quad P|\bar{K}^0\rangle = |\bar{K}^0\rangle, \tag{15.1}$$

$$C|K^0\rangle = |\bar{K}^0\rangle, \tag{15.2}$$

$$T|K^0\rangle = \langle K^0|, \tag{15.3}$$

with similar relations being valid for the antiparticle $|\bar{K}^0\rangle$. Even though $|K^0\rangle$ is a pseudoscalar particle, we chose a convention that under parity $|K^0\rangle$ transforms into itself, since with this choice it is easier to keep track of the minus signs. The freedom to make this choice comes from the fact that parity transformation requires $P^2 = 1$; thus there is still freedom of the overall sign.

The decays of the mesons are mediated by the weak interactions, whose operators are Hermitian. Non-Hermitian terms appear in loop diagrams from energy denominators; however, Hermiticity still holds when we consider the dispersive and absorptive parts separately, as we discuss next. For instance, the semileptonic decay has the amplitude

$$a_\ell = \langle \pi^- \ell^+ \nu | H_\mathrm{W} | K^0 \rangle \tag{15.4}$$

with H_W the weak Lagrangian, which is Hermitian. Similarly, the contributions to the mass matrix, with i and j being K^0 or $\bar{K}^0$, have the general form

$$H_{ij} = m_\mathrm{K}\delta_{ij} + \langle i|H_\mathrm{W}|j\rangle + \sum_n \frac{\langle i|H_\mathrm{W}|n\rangle\langle n|H_\mathrm{W}|j\rangle}{M_\mathrm{K} - E_\mathrm{n} + \mathrm{i}\varepsilon}. \tag{15.5}$$

The first two terms appear for $i = j$. The last term originates from box diagrams and is present for $\Delta S = 2$ transitions with $j = \mathrm{K}^0$ and $i = \bar{\mathrm{K}}^0$. The last term is decomposed into two Hermitian matrices by decomposing the energy denominator into a principal part denoted by P and a δ-function term. The Hamiltonian decomposes as follows:

$$H_{ij} = M_{ij} - \frac{\mathrm{i}}{2}\Gamma_{ij}, \tag{15.6}$$

with a dispersive term

$$M_{ij} = m_\mathrm{K}\delta_{ij} + \langle i|H_\mathrm{W}|j\rangle + P\sum_n \frac{\langle i|H_\mathrm{W}|n\rangle\langle n|H_\mathrm{W}|j\rangle}{M_\mathrm{K} - E_\mathrm{n}} \tag{15.7}$$

and an absorptive term

$$\Gamma_{ij} = 2\pi \sum_n \langle i|H_\mathrm{W}|n\rangle\langle n|H_\mathrm{W}|j\rangle\delta(E_n - M_\mathrm{K}). \tag{15.8}$$

These terms satisfy the Hermiticity relations

$$M_{ij} = M_{ji}^* \quad \text{and} \quad \Gamma_{ij} = \Gamma_{ji}^*. \tag{15.9}$$

More relations follow from CP and CPT invariance. We present the conditions as two theorems.

Theorem 1 *For a Hamilton operator that is CPT-invariant the amplitudes for the decays of particles and antiparticles are the complex conjugates of each other.*

Proof Let us denote the amplitude for K^0 decay by

$$A_I = \langle X_I|H_\mathrm{W}|K^0\rangle. \tag{15.10}$$

Then, using $(CPT)H_W(CPT)^{-1} = H_W$, we obtain

$$\begin{aligned} A_I &= \langle X_I|(CPT)^{-1}H_W(CPT)|K^0\rangle \\ &= \langle \bar{K}^0|H_W|\tilde{X}_I\rangle = \bar{A}_I^*, \end{aligned} \tag{15.11}$$

where $|\tilde{X}_I\rangle = CP|X_I\rangle$, i.e. the CP conjugate state and

$$\bar{A}_I = \langle \tilde{X}_I|H_W|\bar{K}^0\rangle. \tag{15.12}$$

By applying the theorem to the diagonal elements of the mass matrix H_{ij} and using Hermiticity of the dispersive and absorptive parts, we obtain

$$M_{11} = M_{22} \quad \text{and} \quad \Gamma_{11} = \Gamma_{22}. \tag{15.13}$$

This is the statement that CPT invariance demands the equality of masses and widths for particles and antiparticles. When we apply the theorem to off-diagonal elements, we obtain the relation

$$M_{12} = M_{21}^* \quad \text{and} \quad \Gamma_{12} = \Gamma_{21}^*, \tag{15.14}$$

which is not new, but the Hermiticity relations in Eqs. (15.9).

It follows now that the mass matrix in the $|K^0\rangle$ and $|\bar{K}^0\rangle$ has the form

$$M - \frac{i}{2}\Gamma = \begin{pmatrix} M_{11} - \frac{i}{2}\Gamma_{11} & M_{12} - \frac{i}{2}\Gamma_{12} \\ M_{12}^* - \frac{i}{2}\Gamma_{12}^* & M_{11} - \frac{i}{2}\Gamma_{11} \end{pmatrix}. \tag{15.15}$$

As mentioned already, the form of the diagonal elements follows from conservation of the CPT symmetry. We can make them different, thus introducing by hand a violation of CPT invariance, and study the modifications in the lifetimes and other properties of the states.

The presence of the off-diagonal matrix elements implies the mixing of the states K^0 and $\bar{K}^0$. The physical states are a mixture of them, obtained by diagonalizing the mass matrix, which will be presented in the next section. Additional restrictions, which we describe in the next theorem, are introduced by CP symmetry. □

Theorem 2 *For a Hamiltonian that is CP-invariant, the decay amplitudes for particles and antiparticles are relatively real.*

Proof

(i) As before we denote by A_I and $\bar{A}_I$ the decay amplitudes for particles and antiparticles, respectively.

(ii) CP invariance implies

$$H_W = (CP)^{-1}H_W(CP). \tag{15.16}$$

(iii) Then the matrix elements are related,

$$
\begin{aligned}
A_I &= \langle X_I|(CP)^{-1}H_{\mathrm{W}}(CP)|K^0\rangle \\
&= \langle \tilde{X}_I|H_{\mathrm{W}}|\bar{K}^0\rangle = \bar{A}_I .
\end{aligned}
\tag{15.17}
$$

We can apply the theorem in cases in which $|X_I\rangle$ is a specific final state or in the case $|X_i\rangle = |\bar{K}^0\rangle$ which refers to the mass matrix.

Let us consider the latter case first. Taking $|X_I\rangle = |\bar{K}^0\rangle$ and $|\tilde{X}_I\rangle = |K^0\rangle$, we obtain the relation

$$
\langle \bar{K}^0|H_{\mathrm{W}}|K^0\rangle = \langle K^0|H_{\mathrm{W}}|\bar{K}^0\rangle . \tag{15.18}
$$

In this and the following equations H_{W} can be the lowest-order Lagrangian or may include higher-order terms with possible contractions between fields. It follows now that, when we consider the dispersive and absorptive terms separately, they are relatively real. This is a stronger restriction than the Hermiticity requirement of Eq. (15.15), where they were complex conjugates of each other. We shall use these properties in the next section, where we will define the parameter ε.

For decays of particles the theorem says that the amplitudes for particles and antiparticles are relatively real. This indicates a strategy for detecting CP violation. It consists of measuring the phase difference of the two amplitudes relative to a third standard phase, such as the phase occuring in the time development of states, the phase in a Breit–Wigner propagator, or some other known phase. We shall describe several methods in the next sections. □

15.3 Time development of states

In the following sections of Chapter 15, we shall assume that CPT is a good symmetry of Nature and study cases in which the CP symmetry is broken. The fact that there are off-diagonal elements in Eq. (15.5) means that $|K^0\rangle$ and $|\bar{K}^0\rangle$ are not mass eigenstates but the physical states are mixed states. The physical states are obtained by diagonalizing the matrix in Eq. (15.15). Beyond the solution of the physical states we are interested in learning how the elements M_{12} and Γ_{12} are produced. In gauge theories they originate from box diagrams and lead, for the various mesons, to terms of different magnitudes, so that the physical properties of K^0, D^0, and B^0 mesons are very different. We describe first the time development of the states.

We are interested in defining a state that is a superposition of $|K^0\rangle$ and $|\bar{K}^0\rangle$ and has the time development

$$
\psi_1(t) = (B_1|K^0\rangle + D_1|\bar{K}^0\rangle)\mathrm{e}^{\mathrm{i}E_1 t} , \tag{15.19}
$$

with B_1 and D_1 being constants. The time evolution is described by the Schrödinger equation

$$\mathrm{i}\frac{\mathrm{d}}{\mathrm{d}t}\psi_1(t) = \left(M - \mathrm{i}\frac{\Gamma}{2}\right)\psi_1(t), \tag{15.20}$$

whose stationary solutions are determined by the eigenvalue problem

$$\begin{pmatrix} H & H_{12} \\ H_{21} & H \end{pmatrix}\begin{pmatrix} B_1 \\ D_1 \end{pmatrix} = E_1\begin{pmatrix} B_1 \\ D_1 \end{pmatrix}, \tag{15.21}$$

with a similar equation for the second eigenvalue E_2.

The solutions have energies $E_{1,2} = H \pm \sqrt{H_{12}H_{21}}$ and the eigenfunctions

$$\psi_{1,2} = \begin{pmatrix} 1 \\ \pm q/p \end{pmatrix} \mathrm{e}^{-\mathrm{i}E_{1,2}t} \quad \text{with} \quad \frac{q}{p} = \left(\frac{H_{21}}{H_{12}}\right)^{\frac{1}{2}} = \mathrm{e}^{\mathrm{i}\xi}, \tag{15.22}$$

respectively. We have written the wave functions in terms of the parameter q/p and have not yet normalized them. The reason for this is that one frequently uses another parameter, ε, which is defined by

$$\frac{q}{p} = \frac{1-\varepsilon}{1+\varepsilon} = \frac{[M_{12}^* - (\mathrm{i}/2)\Gamma_{12}^*]^{1/2}}{[M_{12} - (\mathrm{i}/2)\Gamma_{12}]^{1/2}} = \mathrm{e}^{\mathrm{i}\xi}, \tag{15.23}$$

which will be used later on. The dynamics of each problem resides in the matrix elements of the Hamiltonian, which are calculated in terms of the box diagrams. We describe the results of these calculations in the next section. They provide us with values for M_{12} and Γ_{12}, which turn out to be complex functions indicating CP violation in the mass matrix.

Let us discuss the physical states. On substituting q/p into Eq. (15.19), we obtain at $t = 0$ two normalized states,

$$|K_\mathrm{S}\rangle = \frac{1}{\sqrt{2}(1+|\varepsilon|^2)^{1/2}}\left[(1+\varepsilon)|K^0\rangle + (1-\varepsilon)|\bar{K}^0\rangle\right], \tag{15.24}$$

$$|K_\mathrm{L}\rangle = \frac{1}{\sqrt{2}(1+|\varepsilon|^2)^{1/2}}\left[(1+\varepsilon)|K^0\rangle - (1-\varepsilon)|\bar{K}^0\rangle\right]. \tag{15.25}$$

Each state has its own time development given by $\mathrm{e}^{-\mathrm{i}E_{\mathrm{S,L}}t}$. The subscripts S and L indicate the short- and long-lived states. When $\varepsilon = 0$, $|K_\mathrm{S}\rangle$ and $|K_\mathrm{L}\rangle$ are even and odd eigenstates of the CP operator. For Re $\varepsilon \neq 0$ the states are no longer CP eigenstates, indicating that the symmetry is broken in the construction of the states.

There are two special properties we wish to discuss. The states are not orthogonal to each other, but have an overlap

$$\langle K_\mathrm{S}|K_\mathrm{L}\rangle = \frac{2\,\mathrm{Re}\,\varepsilon}{1+|\varepsilon|^2}. \tag{15.26}$$

This follows from the fact that the mass matrix, in general, is not Hermitian. The second property occurs when ε is purely imaginary. We can use the states $|K^0\rangle$ and $|\bar{K}^0\rangle$ or new states

$$|\tilde{K}^0\rangle = e^{i\alpha}|K^0\rangle \quad \text{and} \quad |\tilde{\bar{K}}^0\rangle = e^{-i\alpha}|\bar{K}^0\rangle \tag{15.27}$$

rotated by a constant phase α. A purely imaginary ε can be eliminated by the appropriate redefinition of the states in Eqs. (15.24) and (15.25). The real part of ε cannot be eliminated. A real part of ε_K has been established for the K^0-meson system. For B^0 mesons ε_B is, to a good approximation, purely imaginary and there is no CP violation in the construction of the physical states.

For the time development of the states we separate the eigenvalues into their respective dispersive and absorptive parts,

$$M_{S,L} - \frac{i}{2}\Gamma_{S,L} = E_{S,L} = E_{1,2},$$

which define the mass and width differences

$$M_L - M_S - \frac{i}{2}(\Gamma_L - \Gamma_S) = 2\sqrt{H_{12}H_{21}}. \tag{15.28}$$

A general state is a superposition of the physical states K_S and K_L, with constant coefficients $C_{1,2}$ describing how the state was created at time $t = 0$:

$$\psi(t) = C_1 e^{-i[M_S-(i/2)\Gamma_S]t}|K_S\rangle + C_2 e^{-i[M_L-(i/2)\Gamma_L]t}|K_L\rangle. \tag{15.29}$$

The decay of the state proceeds through strangeness-changing couplings, which requires that we rewrite them in terms of $|K^0\rangle$ and $|\bar{K}^0\rangle$. The time development of a state that at time $t = 0$ began as $|K^0\rangle$ is given by

$$\psi_1(t) = N\left[f_+(t)|K^0\rangle + \frac{q}{p} f_-(t)|\bar{K}^0\rangle \right], \tag{15.30}$$

with

$$f_\pm = e^{-i[M_S-(i/2)\Gamma_S]t} \pm e^{-i[M_L-(i/2)\Gamma_L]t}, \tag{15.31}$$

with N a normalization constant. Similarly, a state that starts at $t = 0$ as $|\bar{K}^0\rangle$ has the time development

$$\psi_2(t) = N'\left[f_-(t)|K^0\rangle + \frac{q}{p} f_+(t)|\bar{K}^0\rangle \right]. \tag{15.32}$$

These equations indicate that a state that started as a pure $|K^0\rangle$ will develop in time a $|\bar{K}^0\rangle$ component through the interference of the two terms. The fact that it involves an interference phenomenon makes possible the separation of the amplitudes, as well as determination of the factor q/p.

Let us consider an experiment in which a state $|K^0\rangle$ was created. This state will evolve into a mixture of both K^0 and $\bar{K}^0$. In fact the probabilities of finding at time t the $|K^0\rangle$ and $|\bar{K}^0\rangle$ components are

$$|\langle K^0|K^0(t)\rangle|^2 = |f_+(t)|^2 = \frac{1}{4}\left[e^{-\Gamma_1 t} + e^{-\Gamma_2 t} + 2\cos(\Delta M\, t)e^{-\Gamma t}\right]$$

and

$$|\langle \bar{K}^0|K^0(t)\rangle|^2 = \left|\frac{q}{p} f_-(t)\right|^2 = \frac{1}{4}\left|\frac{q}{p}\right|^2 \left[e^{-\Gamma_1 t} + e^{-\Gamma_2 t} - 2\cos(\Delta M\, t)e^{-\Gamma t}\right],$$

with $\Gamma_1 = \Gamma_S$, $\Gamma_2 = \Gamma_L$, and $\Gamma = \frac{1}{2}(\Gamma_S + \Gamma_L)$. Similar formulas hold for a state that starts as $|\bar{K}^0(t)\rangle$. The detection of $|K^0\rangle$ or $|\bar{K}^0\rangle$ in the final state is carried out by observing their decay products. Thus the final formulas involve an additional amplitude, which introduces its own phase. The time development of the states allows the accurate determination of ΔM and of relative phases. In fact, this property is used heavily in the analysis of experiments.

15.3.1 Simplified formulas for K^0 mesons

Numerous experiments with K-meson beams were able to determine ΔM, $\Delta\Gamma$, and the parameter ε_k. The results suggest several approximations that simplify the equation considerably. For neutral K mesons the mass and width differences are comparable:

$$\Delta M_K = M_L - M_S = 3.52 \times 10^{-15}\,\text{GeV}, \quad \Delta\Gamma_K = \Gamma_L - \Gamma_S = -7.36 \times 10^{-15}\,\text{GeV}\,. \tag{15.33}$$

Measurements in the decays of the particles determine ε_K to be small,

$$|\varepsilon_K| = (2.27 \pm 0.02) \times 10^{-3}, \tag{15.34}$$

with a phase of approximately 45°. At the end of this section we describe an experimental method that determines $|\varepsilon_K|$.

For small ε_K the exponent ξ which occurs in Eq. (15.23) is small and allows the following approximations:

$$H_{12} \approx \sqrt{H_{12}H_{21}}(1 - i\xi),$$
$$H_{21} \approx \sqrt{H_{12}H_{21}}(1 + i\xi).$$

From the definition of ε_K it follows that

$$\varepsilon_K = \frac{H_{12} - H_{21}}{H_{12} + H_{21} + 2\sqrt{H_{12}H_{21}}}$$

and, using the above approximations,

$$\varepsilon_K = \frac{i\,\mathrm{Im}\,M_{12} + \frac{1}{2}\mathrm{Im}\,\Gamma_{12}}{2\sqrt{H_{12}H_{21}}} = \frac{i\,\mathrm{Im}\,M_{12} + \frac{1}{2}\mathrm{Im}\,\Gamma_{12}}{\Delta M - (i/2)\Delta\Gamma}. \tag{15.35}$$

The fact that the phase is 45° means that $\mathrm{Im}\,M_{12}$ and $\mathrm{Im}\,\Gamma_{12}$ are comparable. Furthermore, the magnitude of $|\varepsilon_K|$ implies that the denominator is much larger, i.e.

$$\mathrm{Im}\,M_{12},\ \mathrm{Im}\,\Gamma_{12} \leq \mathrm{Re}\,M_{12} \quad \text{or} \quad \mathrm{Re}\,\Gamma_{12}, \tag{15.36}$$

giving the final result

$$H_{12} = \mathrm{Re}\,M_{12} - \frac{i}{2}\mathrm{Re}\,\Gamma_{12} \tag{15.37}$$

and

$$\sqrt{H_{12}H_{21}} \approx \mathrm{Re}\,M_{12} - \frac{i}{2}\mathrm{Re}\,\Gamma_{12}. \tag{15.38}$$

The simplified formulas of this section hold only for the K^0 system. For this case the mass difference is

$$\Delta M = 2\,\mathrm{Re}\,M_{12} \tag{15.39}$$

and the width difference is

$$\Delta\Gamma = 2\,\mathrm{Re}\,\Gamma_{12}. \tag{15.40}$$

We shall discuss the theoretical determination of these quantities in the next section. Before leaving the discussion of the K mesons, we discuss the measurement of $\mathrm{Re}\,\varepsilon_K$ from semileptonic decays.

Let us consider a beam that consists of K_L mesons. This beam is created in accelerators by producing intense beams of K^0 or $\bar{K}^0$ mesons and setting up an experiment far away from the production region, where the K_S particles have already decayed. Next we consider the decays

$$K_L \to \pi^-\ell^+\nu \tag{15.41}$$

and

$$K_L \to \pi^+\ell^-\bar{\nu}, \tag{15.42}$$

distinguished by the charges of the pions and leptons. We denote the decay amplitudes as

$$a_\ell = \langle \pi^-\ell^+\nu | H_W | K^0 \rangle, \tag{15.43}$$

$$\bar{a}_\ell = \langle \pi^+\ell^-\bar{\nu} | H_W | \bar{K}^0 \rangle. \tag{15.44}$$

Assuming CP invariance for the amplitudes, Theorem 2 says that the two amplitudes are equal,

$$a_\ell = \bar{a}_\ell, \tag{15.45}$$

which, together with the definition of K_L, determines the asymmetry

$$\delta = \frac{\Gamma(K_L \to \pi^-\ell^+\nu) - \Gamma(K_L \to \pi^+\ell^-\bar{\nu})}{\Gamma(K_L \to \pi^-\ell^+\nu) + \Gamma(K_L \to \pi^+\ell^-\bar{\nu})} = \frac{2\,\mathrm{Re}\,\varepsilon}{1+|\varepsilon|^2}. \tag{15.46}$$

The experimental value for the asymmetry is

$$\delta = (3.27 \pm 0.12) \times 10^{-3}, \tag{15.47}$$

which is consistent with the magnitude and phase given earlier in this section. The separation into magnitude and phase is obtained by comparing the semileptonic decay with the $K_L \to \pi\pi$ decays, which we shall describe in a following section.

15.4 The K^0–$\bar{K}^0$ transition amplitude

The theoretical calculations of the matrix elements M_{12} and Γ_{12} are far from being understood. For these as well as other matrix elements, there have been developed several methods that provide acceptable values and even make successful predictions. For K^0 mesons there are short- and long-distance contributions, with the dominance of the short-distance contributions being harder to justify, because the strong coupling constant $\alpha_S(q^2)$ is large and the quarks are confined into hadrons. For the B^0 mesons, on the other hand, the dominance of the top quark in intermediate states makes short-distance dominance more reliable.

The diagonal elements of the mass matrix are created by the strong interactions. The off-diagonal term M_{12} involves a $\Delta S = 2$ transition and receives contributions from the box diagrams, as described in Section 14.7. The method consists of calculating an effective $\Delta S = 2$ Hamiltonian in the free-quark model and then taking its matrix element between the K^0 and $\bar{K}^0$ states. The $\Delta S = 2$ Hamiltonian generated by this method was described in Section 14.7:

$$H_W^{\Delta S=2} = -\frac{G^2}{16\pi^2} M_W^2 Q_{\Delta S=2}\left[\lambda_c^2 E(x_c) + 2\lambda_c\lambda_t E(x_c,\, x_t) + \lambda_t^2 E(x_t)\right], \tag{15.48}$$

with the various terms defined as follows. The variable $x_i = m_i^2/M_W^2$ and the couplings of the quarks at the various vertices produce the factors

$$\lambda_i = V_{id}^* V_{is}. \tag{15.49}$$

Their numerical values are determined by the CKM-matrix elements with λ_c being of $O(\lambda)$ and λ_t of $O(\lambda^5)$ in the Wolfenstein parametrization. The functions $E(x_c)$

and $E(x_c, x_t)$ are obtained from the integration over the loop. We wrote down $E(x_i)$ in Eq. (14.52) and the second function is given as

$$E(x_i, x_j) = -x_i x_j \left[\frac{1}{x_i - x_j} \left(\frac{1}{4} + \frac{3}{2}\frac{1}{(1-x_i)} - \frac{3}{4}\frac{1}{(1-x_i)^2} \ln x_i \right) + (i \leftrightarrow j) \right.$$
$$\left. - \frac{3}{4}\frac{1}{(1-x_i)(1-x_j)} \right]. \qquad (15.50)$$

For

$$m_c^2 \ll M_W^2, \quad E(x_c) \to -x_c$$

and for

$$m_t^2 \gg M_W^2, \quad E(x_t) \to -\frac{1}{4}x_t + \frac{3}{2}\ln x_t,$$

which indicates that the various terms in (15.50) are comparable.

Finally, there is the operator

$$Q_{\Delta S=2} = \bar{d}\gamma_\mu(1-\gamma_5)s\bar{d}\gamma^\mu(1-\gamma_5)s,$$

which represents the external lines of the box diagram. The matrix element $X_K = \langle K^0 | Q_{\Delta S=2} | \bar{K}^0 \rangle$ contains the long-distance contribution of this calculation. A good deal of effort has been invested in its calculation. In various situations so far, we have encountered the calculation of two-quark operators (currents) between hadronic states, for which there are reliable numerical estimates – sometimes extracted from experimental data. Estimates of matrix elements for four-quark operators are less reliable and are still a subject of research. A simple estimate of such matrix elements is given in Eq. (16.5), which can be taken over for the K mesons by making the replacements $F_D \to F_K$ and $M_D \to M_K$.

The absorptive part Γ_{12} is in principle also calculable in terms of the box diagrams by setting the intermediate states on the mass shell, i.e. replacing the propagators by δ-functions. For K mesons the physical intermediate states are u quarks, making the absorptive part a long-distance effect. This term is calculated by low-energy methods with the intermediate states being 2π, 3π, ... mesons. The calculation carries a large uncertainty because the amplitudes and their relative phases are not known.

The situation is different for heavy mesons, in particular the B^0–$\bar{B}^0$ system, in which there are many intermediate states with multiparticle final states dominating the decay. For heavy mesons the sum over intermediate states will be replaced by the quarks and will be calculated as the absorptive part of the diagram. This is known as the quark–hadron duality, whereby hadronic matrix elements are replaced by the corresponding quark diagrams. Finally, the matrix elements are taken between hadrons, for which approximations are again necessary.

We close this section by deriving two approximate formulas describing the mass and width differences of neutral B mesons. The general formulas for width and mass differences are

$$\Delta M = 2\,\mathrm{Re}\left[\left(M_{12} - \frac{i}{2}\Gamma_{12}\right)\left(M_{12}^* - \frac{i}{2}\Gamma_{12}^*\right)\right]^{\frac{1}{2}}, \tag{15.51}$$

$$\Delta\Gamma = -4\,\mathrm{Im}\left[\left(M_{12} - \frac{i}{2}\Gamma_{12}\right)\left(M_{12}^* - \frac{i}{2}\Gamma_{12}^*\right)\right]^{\frac{1}{2}}. \tag{15.52}$$

For these equations, we need assume only that the CPT symmetry is exact. For K^0 mesons the data imply the approximations which were described in Section 15.3.1. For the B mesons the situation is different. Estimates of Γ_{12} and M_{12} using the effective Hamiltonians of this section lead to the estimate (Paschos and Türke, 1989)

$$\Gamma_{12} \sim 0.1 M_{12}$$

and with almost the same phase; consequently for B mesons

$$\Delta M_B = 2|M_{12}| \quad \text{and} \quad \Delta\Gamma_B = 2|\Gamma_{12}| \tag{15.53}$$

to a good approximation. The differences in the masses and widths of the K^0 and B^0 mesons indicate that each system must be treated separately. The qualitative differences are understood in terms of the quark substructure which enters the box diagrams.

15.5 CP violation in amplitudes

Besides the phase introduced in the mass matrix the decay amplitudes have their own phases. Theorem 1 states that the amplitudes for particle and antiparticle decays are the complex conjugates of each other. This is a consequence of CPT. CP symmetry goes one step further and requires the amplitudes to be real relative to each other. Consequently, evidence for the breakdown of CP symmetry requires the measurement of phases.

In quantum mechanics the overall phase of a sum of amplitudes can always be removed, but relative phases among amplitudes are measurable observables. For this reason all measurements must include at least two phases and the experiments measure one phase relative to the other.

Let us denote the final state by $\langle f|$ and in addition select the final state to be a CP eigenstate with eigenvalue unity. Examples of such decays are

$$\mathrm{K}^0 \to \pi^+\pi^-, \;\; \pi^0\pi^0. \tag{15.54}$$

There are two decay amplitudes specified by the isospin of the two pions being 0 or 2. Searches for direct CP violation measure the relative phase of the two amplitudes and try to establish whether it is the same in K^0 and $\bar{\mathrm{K}}^0$ decays. We denote the amplitude

$$\begin{aligned} \langle(2\pi)_I|H_\mathrm{W}|K^0\rangle &= A_I \mathrm{e}^{\mathrm{i}\delta_I}, \\ \langle(2\pi)_I|H_\mathrm{W}|\bar{K}^0\rangle &= \bar{A}_I \mathrm{e}^{\mathrm{i}\delta_I} \quad \text{with} \quad I = 0 \text{ or } 2. \end{aligned} \tag{15.55}$$

The phases δ_I are created by final-state interactions of the two pions, which is a strong-interaction effect independent of the initial state but a function of the isospin I. Beyond the strong phase there is also a phase of weak origin, which changes sign as we go from particles to antiparticles. Consequently, we can write the amplitudes as follows.

$$A_I = |A_I|\mathrm{e}^{\mathrm{i}\theta_I}, \tag{15.56}$$

$$\bar{A}_I = |A_I|\mathrm{e}^{-\mathrm{i}\theta_I}, \tag{15.57}$$

where θ_I is now a phase of weak origin.

Experiments starting with a $|K^0\rangle$ or a $|\bar{K}^0\rangle$ beam also observed the mixing phenomenon, described in Section 15.3, in the decays to $\pi^+\pi^-$ and $\pi^0\pi^0$. At a distance corresponding to six to seven lifetimes of the $\mathrm{K_S}$ mesons the two amplitudes interfere and show a difference. In this way one can separate the ratios

$$\eta_{+-} = \frac{A(\mathrm{K_L} \to \pi^+\pi^-)}{A(\mathrm{K_S} \to \pi^+\pi^-)} \tag{15.58}$$

and

$$\eta_{00} = \frac{A(\mathrm{K_L} \to \pi^0\pi^0)}{A(\mathrm{K_S} \to \pi^0\pi^0)}. \tag{15.59}$$

If CP is a good symmetry (the CP quantum number is conserved), then these ratios vanish. The experiments found these ratios to be different from zero. It is customary to make an isospin analysis of the amplitudes and write them as

$$A(\mathrm{K}^0 \to \pi^0\pi^0) = \sqrt{\frac{2}{3}}A_0 - \frac{2}{\sqrt{3}}A_2 \tag{15.60}$$

and

$$A(\mathrm{K}^0 \to \pi^+\pi^-) = \sqrt{\frac{2}{3}}A_0 + \frac{1}{\sqrt{3}}A_2 \tag{15.61}$$

and rewrite the ratios in terms of isospin amplitudes.

Straightforward substitution of the amplitudes gives

$$\eta_{+-} = \frac{(\sqrt{2}A_0 + A_2\mathrm{e}^{\mathrm{i}\Delta}) - \mathrm{e}^{\mathrm{i}\xi}(\sqrt{2}A_0^* + A_2^*\mathrm{e}^{\mathrm{i}\Delta})}{(\sqrt{2}A_0 + A_2\mathrm{e}^{\mathrm{i}\Delta}) + \mathrm{e}^{\mathrm{i}\xi}(\sqrt{2}A_0^* + A_2^*\mathrm{e}^{\mathrm{i}\Delta})}, \tag{15.62}$$

with $\Delta = \delta_2 - \delta_0$. It is mentioned here that the ratio is phase-convention independent. A popular phase convention was introduced by Wu and Yang, which selects the A_0 amplitude to be real and then the answers will depend on the phase of the A_2 amplitude denoted by θ_2. We adopt this convention; then, by substituting $\mathrm{e}^{\mathrm{i}\xi}$ in terms of ε and collecting terms together, we obtain

$$\eta_{+-} = \frac{\varepsilon\left[1 + \dfrac{1}{\sqrt{2}}\left|\dfrac{A_2}{A_0}\right|\cos\theta_2\,\mathrm{e}^{\mathrm{i}\Delta} + \left(\dfrac{\mathrm{i}}{\sqrt{2}}\right)\left|\dfrac{A_2}{A_0}\right|\sin\theta_2\,\mathrm{e}^{\mathrm{i}\Delta}\right]}{\left[1 + \dfrac{1}{\sqrt{2}}\left|\dfrac{A_2}{A_0}\right|\cos\theta_2\,\mathrm{e}^{\mathrm{i}\Delta} + \left(\dfrac{\mathrm{i}\varepsilon}{\sqrt{2}}\right)\left|\dfrac{A_2}{A_0}\right|\sin\theta_2\,\mathrm{e}^{\mathrm{i}\Delta}\right]}. \tag{15.63}$$

The expression simplifies if we neglect the second term in the denominator, since $\varepsilon|A_2/A_0|\sin\theta_2 \ll 1$. In this case

$$\eta_{+-} = \varepsilon + \frac{\varepsilon'}{1 + \omega/\sqrt{2}} \tag{15.64}$$

with

$$\varepsilon' = \left(\frac{\mathrm{i}}{\sqrt{2}}\right)\left|\frac{A_2}{A_0}\right|\sin\theta_2\,\mathrm{e}^{\mathrm{i}\Delta}$$

and

$$\omega = \left|\frac{A_2}{A_0}\right|\cos\theta_2\,\mathrm{e}^{\mathrm{i}\Delta}.$$

On repeating the analysis for η_{00} with the same approximations, we obtain

$$\eta_{00} = \varepsilon - \frac{2\varepsilon'}{1 - \sqrt{2}\,\omega}. \tag{15.65}$$

In many models both A_0 and A_2 are complex and rephasing of the amplitudes is necessary in order to bring them into accord with the Wu–Yang phase convention.

In summary, in addition to the CP parameter discussed in Section 15.3.1, there is the parameter ε'. The parameter ε arose from phases in the mass matrix and ε' from relative phases in the decay amplitudes. The former is referred to as indirect

CP violation and the latter as direct CP violation. It is also customary to define the ratio ε'/ε because the phase $\pi/2 - \Delta \approx 45^\circ$ of ε' is approximately equal to the phase of ε and cancels out in the ratio.

Going back to a general phase convention whereby both A_0 and A_2 are complex, we should replace θ_2 by $\theta_2 - \theta_0$ and the definition of ε'/ε becomes

$$\begin{aligned}\frac{\varepsilon'}{\varepsilon} &= \frac{1}{\sqrt{2}|\varepsilon|}\left|\frac{A_2}{A_0}\right|(\sin\theta_2 - \sin\theta_0)\\ &= -\frac{\omega}{\sqrt{2}\,\varepsilon}\frac{1}{\text{Re}\,A_0}\left(\text{Im}\,A_0 - \frac{1}{\omega}\text{Im}\,A_2\right),\end{aligned} \tag{15.66}$$

with

$$\omega = \left|\frac{A_2}{A_0}\right| \approx \frac{\text{Re}\,A_2}{\text{Re}\,A_0} \simeq \frac{1}{22.2}.$$

The remaining problem is the calculation of the imaginary parts of the amplitudes in Eq. (15.66) because the real parts are much larger, with their numerical values known from experiments.

15.6 The effective Hamiltonian

K-meson decays involve low-energy interactions mediated by the exchanges of hadrons and at least one W boson. It is customary to appeal to the quark–hadron duality and replace the hadrons by quarks and gluons. The weak interaction is a short-distance phenomenon that is represented by the couplings of the W to quarks. This is not the only part of the interaction, because there are strong interactions produced by the exchanges of gluons. A complete calculation must include both of them. Thus a method has been developed in field theory for this purpose. The method consists of summing the leading logarithmic contributions of the diagrams. The final result is an effective field theory with the W and the heavy quarks eliminated or, as one says, "they have been integrated out."

Even if we start with one weak operator at momenta comparable to M_W, the exchange of gluons introduces more operators coming from loop diagrams, like penguin and box diagrams. The effective Hamiltonian has the form

$$H_{\text{eff}} = \sum_{a,b} C_{ab}(M_W,\, \mu)Q_{ab}(\mu), \tag{15.67}$$

with $Q_{ab}(\mu) = \bar{q}(x)\Gamma_a q(x)\bar{q}(x)\Gamma_b q(x)$ with Γ_a and Γ_b being matrices in Dirac space. The constants $C_{ab}(M_W,\, \mu)$ are the coefficient functions (Wilson coefficients) obtained from the renormalization of the operators. They depend on a high energy,

M_W, a low energy scale, μ, and the quarks in the intermediate states. Their general form is

$$C_{ij}(M_W, \mu) \sim \ln\left(\frac{M_W}{\mu}\right)^{\frac{\gamma}{b}}, \tag{15.68}$$

which is obtained by integrating renormalization equations of quantum chromodynamics (QCD). The exponent γ is known as the anomalous dimension and b arises from the running of the coupling constant. The calculation of the coefficients and their accuracy is a theoretical topic of active research whose study is beyond the scope of this book (Buchalla *etal.*, 1996).

In order to give a general impression of the results, we present here the effective Hamiltonian for K-meson decays. As mentioned already, it depends only on the light quarks and contains eight operators:

$$\mathcal{H}_{\text{eff}}^{\Delta S=1} = \frac{G_F}{\sqrt{2}} \sum_{i=1}^{8} \left[C_i^u(\mu)\lambda_u + C_i^c(\mu)\lambda_c + C_i^t(\mu)\lambda_t\right] Q_i \quad \text{for} \quad \mu < m_c, \tag{15.69}$$

where $\lambda_q = V_{qd}^* V_{qs}$ are again the couplings from the CKM matrix, the unitarity of which implies

$$\lambda_u + \lambda_c + \lambda_t = 0 \tag{15.70}$$

and makes possible the elimination of one of them. Once we decide to eliminate λ_u, the coefficient functions will appear as differences $C_i^c - C_i^u$ and $C_i^t - C_i^u$. The substitution makes the coefficient functions less sensitive to the up quark. The operators which appear are defined as follows:

$$\begin{aligned}
Q_1 &= 4\bar{s}_L\gamma^\mu d_L \bar{u}_L\gamma_\mu u_L, & Q_2 &= 4\bar{s}_L\gamma^\mu u_L\, \bar{u}_L\gamma_\mu d_L,\\
Q_3 &= 4\sum_q \bar{s}_L\gamma^\mu d_L\, \bar{q}_L\gamma_\mu q_L, & Q_4 &= 4\sum_q \bar{s}_L\gamma^\mu q_L\, \bar{q}_L\gamma_\mu d_L,\\
Q_5 &= 4\sum_q \bar{s}_L\gamma^\mu d_L\, \bar{q}_R\gamma_\mu q_R, & Q_6 &= -8\sum_q \bar{s}_L q_R\, \bar{q}_R d_L,\\
Q_7 &= 4\sum_q \frac{3}{2} e_q \bar{s}_L\gamma^\mu d_L\, \bar{q}_R\gamma_\mu q_R, & Q_8 &= -8\sum_q \frac{3}{2} e_q \bar{s}_L q_R\, \bar{q}_R d_L.
\end{aligned} \tag{15.71}$$

Operator Q_2 is the original charged-current operator and Q_1 is generated from box-type diagrams, where in addition to the W a gluon is being exchanged (Fig. 15.1).

The penguin diagrams in Fig. 15.2 generate $Q_3, \ldots, Q_6$. Finally, penguin diagrams with the exchange of photons generate Q_7 and Q_8 (electroweak penguins). Since the penguin diagrams are important, we present several steps of the calculation in Section 15.7, where it is also explained how the penguin diagrams generate the various operators.

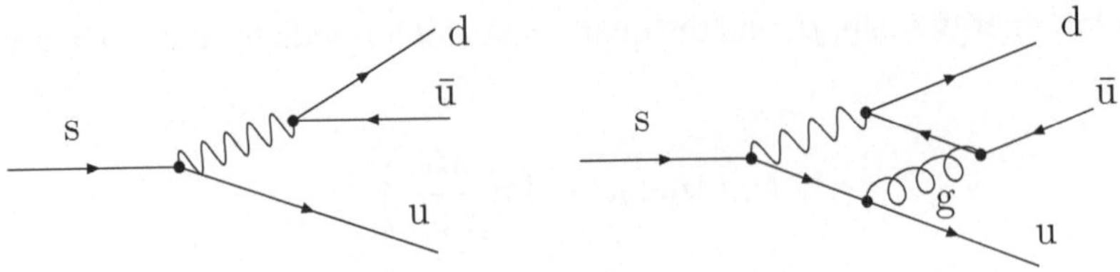

Figure 15.1. Tree and box diagrams.

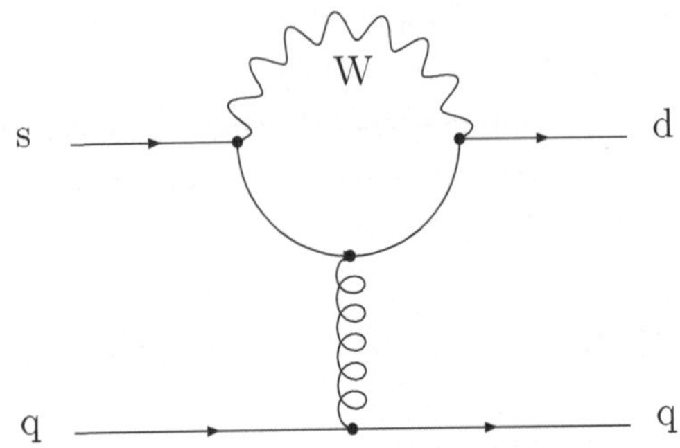

Figure 15.2. A penguin diagram.

Finally, one may also show that not all the operators are independent, since they satisfy the relation

$$-Q_1 + Q_2 + Q_3 = Q_4.$$

The coefficient functions $C_i(M_\mathrm{W}, \mu)$ originate from the short-distance interaction of QCD and have been calculated at the one-loop, as well as the two-loop, level. The hadronic matrix elements

$$\langle Q_i(\mu)\rangle_I = \langle \pi\pi, I|Q_i(\mu)|K^0\rangle \tag{15.72}$$

originate from long-distance interactions, since they involve low energies and momenta. They represent the low-energy limit of QCD and must be calculated by low-energy methods. They have been the subject of various calculations, which we shall mention briefly. Within the framework described in this section we can outline the calculation of ε'/ε.

Among the amplitudes which enter the calculation, Re A_0 and Re A_2 are taken from experimental data (Devlin and Dickey, 1979):

$$\mathrm{Re}\, A_0 = 0.338 \times 10^{-6}\,\mathrm{GeV} \quad \text{and} \quad \mathrm{Re}\, A_2 = 0.015 \times 10^{-6}\,\mathrm{GeV}.$$

They are much larger than their imaginary parts. The amplitudes Im A_0 and Im A_2 are calculated in terms of the effective Hamiltonian

$$\mathrm{Im}\,A_I = \frac{G}{\sqrt{2}} \sum_{i=1}^{8} [(C_i^{\mathrm{c}} - C_i^{\mathrm{u}})\mathrm{Im}\,\lambda_{\mathrm{c}} + (C_i^{\mathrm{t}} - C_i^{\mathrm{u}})\mathrm{Im}\,\lambda_{\mathrm{t}}]\langle Q_i\rangle_I. \tag{15.73}$$

With a phase convention of the CKM matrix whereby V_{us} and V_{ud} are real, the unitarity of the CKM matrix provides one additional relation,

$$\mathrm{Im}\,\lambda_{\mathrm{t}} = -\mathrm{Im}\,\lambda_{\mathrm{c}} = \lambda^5 A\eta. \tag{15.74}$$

Substitution of this relation eliminates the u quarks in intermediate states, since the Wilson coefficients for top and charm quarks are subtracted from each other. This leads to the result

$$\frac{\varepsilon'}{\varepsilon} = \frac{G_{\mathrm{F}}}{\sqrt{2}} \frac{\omega}{|\varepsilon|} \frac{1}{\mathrm{Re}\,A_0} \mathrm{Im}\,\lambda_{\mathrm{t}} \sum y_i^{\mathrm{t}}(\mu) \left[\langle Q_i(\mu)\rangle_0 - \frac{1}{\omega} \langle Q_i(\mu)\rangle_2 \right], \tag{15.75}$$

with $y_i^{\mathrm{t}}(\mu) = C_i^{\mathrm{t}}(\mu) - C_i^{\mathrm{c}}(\mu)$ and $\omega = \mathrm{Re}\,A_2/\mathrm{Re}\,A_0$. The superscripts denote contributions from top and charm quarks in the intermediate states. The unitarity of the CKM matrix helps by eliminating the Wilson coefficients of the u quarks and making the QCD contribution sensitive to the energy scales between m_{c} and m_{t}, where the short-distance expansion is acceptable. The Wilson coefficients are available and have been tabulated (Buchalla *et al.*, 1996).

The hadronic matrix elements have been the subject of numerous calculations. From the early estimates it was evident that $\langle Q_6\rangle_0$ plays an important role. The matrix element is generated by the penguin diagrams and, since it involves pseudoscalar densities, it is enhanced. In chiral perturbation theory it is expressed in terms of coupling constants divided by the mass of the strange quark. It was also calculated by vacuum saturation or the tree contribution of the chiral perturbation theory. It was noted that the lowest-order contribution must be supplemented by chiral loops (Bardeen *et al.*, 1987, 1998). The final results indicate that $\langle Q_6\rangle_0$ is important, especially because it is further enhanced by contributions from chiral loops.

An additional complication arises from the matrix element $\langle Q_8\rangle_2$, whereby in the penguin diagrams the gluon is replaced by a photon. It was argued that the electroweak term can be very important because it is multiplied by a large factor, $1/\omega$. Calculations in chiral perturbation theory indicate that its contribution is moderate and it is further reduced by loops. A good approximation consists of taking the dominance $\langle Q_6\rangle_0$ and $\langle Q_8\rangle_2$. As an illustrative example we give typical values for the Wilson coefficients,

$$y_6 = -0.110 \quad \text{and} \quad y_8 = 1.15 \times 10^{-3}, \tag{15.76}$$

and the matrix elements,

$$\langle Q_6\rangle_0 = -3.4\,\mathrm{GeV}^3 \qquad \text{and} \qquad \langle Q_8\rangle_2 = 0.46\,\mathrm{GeV}^3, \tag{15.77}$$

obtained in chiral perturbation theory for $m_s = 150$ MeV. The CKM term is precisely known to be

$$\mathrm{Im}\,\lambda_t = (1.35 \pm 0.35) \times 10^{-3}. \tag{15.78}$$

On collecting all terms together in Eq. (15.75), we obtain the value

$$\left(\frac{\varepsilon'}{\varepsilon}\right)_K = 16.9 \times 10^{-4}, \tag{15.79}$$

which is consistent with the experimental values. The contribution of the electroweak penguin is less than 20%. Many quantities entering the calculation carry uncertainties and the final range for the ratio is larger, in the range $(10\text{–}20) \times 10^{-4}$ (Hambye *et al.*, 2000). Calculations in the chiral quark model give a similar range. These values are consistent with the newest experimental values,

$$\left(\frac{\varepsilon'}{\varepsilon}\right)_K = \begin{cases} (14.7 \pm 2.2) \times 10^{-4} & \text{NA(48) (Fanti } et\ al.\text{, 1999),} \quad (15.80) \\ (22.7 \pm 2.8) \times 10^{-4} & \text{KTeV (Alavi-Harati } et\ al.\text{, 1999).} \quad (15.81) \end{cases}$$

At this time it seems that the experiments are more precise than the theory. This is the outcome of four large experiments that invested great efforts in measuring precisely decays and interference phenomena in K-meson decays.

It would be an omission not to mention a good deal of work done on lattice gauge theories, which tries to determine the matrix elements. Unfortunately, their results are not stable enough yet. They give a wide range of values for the matrix elements and the CP parameter.

This is an introduction to the calculations of the CP parameter intended for students who may use it as a guide to the published articles. The bottom line is that theoretical analyses in the standard model are consistent with experimental measurements. The CKM paradigm gives a consistent – albeit not very accurate – picture for the K-meson decays and it remains to find out whether it continues being successful for mesons containing heavy quarks.

15.7 Calculation of a penguin diagram

In K-meson and B-meson decays an important contribution comes from the penguin diagram. We have mentioned already that in Eq. (15.71) the penguin diagram with gluonic corrections produces four operators. It is worthwhile to give several steps of the calculation, which demonstrate how the various operators are generated. This

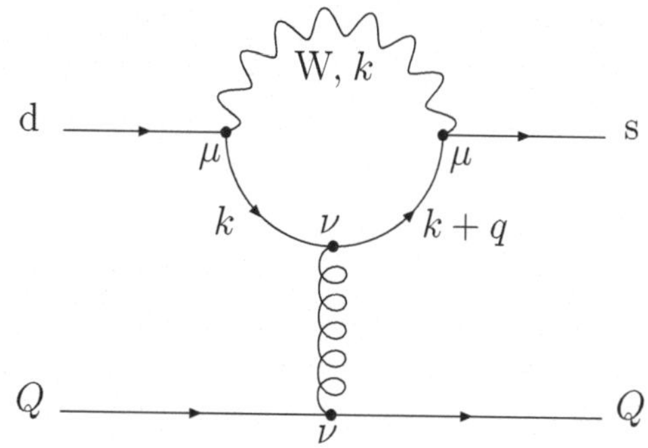

Figure 15.3. Momentum assignments for the penguin diagram.

section contains long algebraic manipulations and is presented here for those who are theoretically inclined.

The notation for the penguin diagram is introduced in Fig. 15.3. The external momenta are those of the strange and down, quarks, which correspond to typical momenta within light mesons and are small relative to the mass of the W boson. For this reason external momenta are kept in the spinors but will be neglected within the four-dimensional integral. A peculiarity of the penguin diagram is the presence of the gluon propagator with momentum q, which is kept throughout the calculation. Following standard rules, the matrix element is written in the form

$$\mathcal{M}_{\mathrm{P}} = \frac{g_{\mathrm{W}}^2}{8}\frac{g_{\mathrm{s}}^2}{q^2}\int \bar{s}\gamma_\mu\gamma_- \frac{\not{k}+\not{q}+m_i}{(k+q)^2-m_i^2}\gamma_\nu \frac{\lambda^\alpha}{2}\frac{\not{k}+m_i}{k^2-m_i^2}\gamma^\mu\gamma_- d\,\frac{1}{k^2-M_{\mathrm{W}}^2}\frac{\mathrm{d}^4k}{(2\pi)^4}$$
$$\times\,\bar{Q}\gamma^\nu\frac{\lambda^\alpha}{2}Q\cdot V_{is}^* V_{id}. \qquad (15.82)$$

The quarks in the loop and their masses are denoted by the subscript i and m_i, respectively. The index for the intermediate quarks occurs also in the CKM matrix elements V_{is} and V_{id}. The rest of the notation is standard, with g_{W} and g_{s} the weak and strong coupling constants, respectively, $\gamma_- = (1-\gamma_5)$, and λ^α the color matrices.

We follow several of the steps for the calculation of loops described in Section 14.7.2. We rewrite the matrix element as

$$\mathcal{M}_{\mathrm{P}} = \frac{g_{\mathrm{W}}^2}{8}\frac{g_{\mathrm{s}}^2}{q^2}\bar{s}\gamma_\mu\gamma_-\gamma_\alpha\gamma_\nu\left(\frac{\lambda^\alpha}{2}\right)\gamma_\beta\gamma^\mu\gamma_- d\bar{Q}\left(\frac{\lambda^\alpha}{2}\right)QI^{\alpha\beta}(m_i,q), \qquad (15.83)$$

with

$$I_{\alpha\beta}(m_i,m_i,q) = \int\frac{\mathrm{d}^4k}{(2\pi)^2}\frac{(k+q)_\alpha k_\beta}{\left[(k+q)^2-m_i^2\right]\left(k^2-m_i^2\right)\left(k^2-M_{\mathrm{W}}^2\right)}V_{is}^*V_{id}. \qquad (15.84)$$

In this way we have separated the spinor structure from the four-dimensional integral. We ignored fermion masses in the numerator, since they are small relative to the integration momenta, and set $\bar{s}\not{q}d = 0$ because we consider masses for external quarks that are small. There is a logarithmic divergence independent of the quark masses. It is multiplied by $\sum_i V_{is}^* V_{is} = 0$ and vanishes. Similarly, in the limit in which the gluon momentum transfer q^2 is much larger than all internal quark masses, it is easy to check that the penguin diagram vanishes. However, for problems involving $q^2 \ll m_c^2$ or m_t^2 the cancellation is not complete.

The remaining two integrals appear with the factors $k_\alpha k_\beta$ or $q_\alpha k_\beta$ in the numerator. Using the method of Feynman parameters and the integrations described in Problem 4, the dominant contribution of the integral for $m_c,\ m_t \ll M_W$ is

$$I_{\alpha\beta}(m_i,\, q) = -\frac{q_\alpha q_\beta}{M_W^2}\frac{1}{16\pi^2 i}\ln\left(\frac{M_W^2}{m_i^2}\right)\left(\frac{1}{3}-\frac{1}{2}\right)V_{is}^* V_{id}, \tag{15.85}$$

with the $\frac{1}{3}$ coming from the $k_\alpha k_\beta$ term and the $\frac{1}{2}$ from the $q_\alpha k_\beta$ term. Finally, we simplify the spin structure by using known identities:

$$q^\alpha q^\beta \bar{s}\gamma_\beta\gamma_\nu\gamma_\alpha\gamma_- d = \bar{s}(2q_\nu - \gamma_\nu\not{q})\not{q}\gamma_- d = -q^2\bar{s}\gamma_\nu\gamma_- d. \tag{15.86}$$

The q^2 factor cancels out the gluon propagator in Eq. (15.82).

On collecting terms together, we arrive at the final result

$$\mathcal{M}_P = -\frac{G}{\sqrt{2}}\frac{\alpha_s}{12\pi}\ln\left(\frac{M_W^2}{m_i^2}\right)\bar{s}_L\gamma_\nu\lambda^\alpha d_L\bar{Q}\gamma^\nu\lambda^a Q(V_{is}^* V_{id}). \tag{15.87}$$

We note that the coupling $\bar{s}_L \ldots d_L$ contains left-handed quarks, in contrast to the gluon coupling $\bar{Q}\gamma_\nu Q$ being a vector. By decomposing the latter into left- and right-handed couplings, we generate two distinct operators. Finally, using an identity for the product of color matrices,

$$\sum_a \lambda_{ij}^a\lambda_{kl}^a = 2\left(\delta_{il}\delta_{jk} - \frac{1}{3}\delta_{ij}\delta_{kl}\right), \tag{15.88}$$

we double the number of operators. In the end, the penguin diagram generates four operators, Q_3, Q_4, Q_5, and Q_6, which were absent at the tree level.

There are two ways to treat the penguin diagram. One of them considers its contribution as a short-distance operator creating a four-fermion interaction among the quarks. This would be the case when the top quark dominates a process. The exchange of additional gluons may still be soft and some sort of summation is again necessary. A final step is the estimation of the four-quark operator between hadronic states.

The alternative method considers the four operators generated by the penguin diagrams as basic operators and sums up higher-order QCD corrections. This is achieved by considering gluonic corrections to each of the operators $Q_1, \ldots, Q_6$, which renormalizes and in addition mixes them up; that is, gluonic corrections to one operator generate several of the others. The problem to be solved is one of coupled differential equations. The initial conditions are defined at high momenta when only Q_2 has an initial value and all other operators are zero. Following this method (Peskin and Schroeder, 1995; Buchalla *et al.*, 1996), one arrives at the effective Hamiltonian similar to that in Eq. (15.69). The theory is effective because the additional corrections are proportional to higher powers of $\alpha_s(p^2)$, which for large momenta become very small.

Problems for Chapter 15

1. Introduce in Eq. (15.62) the Wu–Yang phase convention, then substitute for $e^{i\xi}$ in terms of ε and derive Eqs. (15.63) and (15.64).
2. (i) The normalizations that occur in Eqs. (15.30) and (15.32) describe how many K^0 or $\bar{K}^0$ mesons, respectively, are present at time $t = 0$. Argue that for normalized wave functions $|K^0\rangle$ and $|\bar{K}^0\rangle$ they should be $N = N' = \frac{1}{2}$.
   (ii) Consider the time development of the state $|K^0\rangle$ and the decays to $\pi^+\pi^-$. Describe the interference term and find an argument justifying the large interference at a distance corresponding to six to seven lifetimes of the K_S meson.
3. Show that the operators Q_1, Q_2, Q_3, and Q_4 satisfy the relation
$$-Q_1 + Q_2 + Q_3 = Q_4.$$
4. The calculation of the integral in Eq. (15.84) contains in the numerator two terms: one with $k_\alpha k_\beta$ and the other with $q_\alpha k_\beta$. Write each of the integrals in terms of Feynman parameters. The four-dimensional integrations are of the form
$$\int \frac{d^4k}{(2\pi)^4} \frac{\{k_\alpha, k_\alpha k_\beta\}}{(k^2 + 2k\cdot p - \Delta)^3} = \frac{1}{32\pi^2 i}\left[\frac{\{-p_\alpha;\ p_\alpha p_\beta\}}{\Delta + p^2 + i\varepsilon} + \left\{0;\ \frac{1}{2} g_{\alpha\beta}\ln(\Delta + p^2 + i\varepsilon) + A_0\right\}\right].$$
The function Δ contains masses of the quarks, the mass M_W, q^2, and Feynman parameters. The constant A_0 is cut-off-dependent but independent of quark masses; it disappears when we sum over the quarks in the loop. The remaining two integrations are elementary. Arrange the integrations in an appropriate way to extract the leading $\ln(M_W/m_i)$ term and obtain Eq. (15.85).

Comment We described the integrals in the limit $m_i \ll M_W$. For $m_t > M_W$ you can again study the elementary integrals and obtain a modified logarithmic term.

References

A. Alavi-Harati, Albuquerque, I. F., Alexopoulos, T., *et al.*, kTeV collaboration (1999), *Phys. Rev. Lett.* **83**, 22
Bardeen, W. A., Buras, A. J., and Gerard, J.-M. (1987), *Nucl. Phys.* **B293**, 787
Bardeen, W. A., Hambye, T., Köhler, G. O., *et al.* (1998), *Phys. Rev.* **D58**, 014017

Buchalla, G., Buras, A., and Lauterbacher, M. (1996), *Rev. Mod. Phys.* **68**, 1125–1199
Devlin, T. J., and Dickey, J. O. (1979), *Rev. Mod. Phys.* **51**, 237
V. Fanti, Lai, A., Marras, D., *et al.*, NA(48) collaboration (1999), *Phys. Lett.* **B465**, 335
Hambye, T., Köhler, G. O., Paschos, E. A., and Soldan, P. H. (2000), *Nucl. Phys.* **B564**, 391
Paschos, E. A., and Türke, U. (1989), *Phys. Rep.* **178**, 145–260
Peskin, M. E., and Schroeder, D. V. (1995), *Quantum Field Theory* (Reading, MA, Perseus Books), especially Chapter 18

Select bibliography

There are several books in which CP properties are discussed. The references given here also useful for Chapter 16.

Bigi, I., and Sanda, A. (2000), *CP Violation* (Cambridge, Cambridge University Press)
Branco, G. C., Lavoura, L., and Silva, J. P. (1999), *CP Violation* (Oxford, Oxford University Press)
Kleinknecht, K. (2003), *Uncovering CP Violation (Experimental Classification in the Neutral* K *Meson and* B *Meson Systems)* (Springer Tracts in Modern Physics) (Berlin, Springer-Verlag)
Lee, T. D. (1981), *Particle Physics and Introduction to Field Theory* (New York, Harwood Academic Publishers)

16

CP violation: D and B mesons

16.1 Introduction

In previous chapters we mentioned that the analysis of bound states with heavy quarks varies from meson to meson. For heavy mesons the hadronic structure becomes simpler and approaches the spectator model with corrections given by HQET. Their weak properties, on the other hand, remain distinct because they depend on the interplay between CKM couplings and the masses of the mesons. For this reason we discuss the mixing and CP violation in D and B mesons in a separate chapter.

The system D^0–$\bar{D}^0$ is quite different because the decay width is much larger than the mass and width differences, which makes the observation of mixing and of CP asymmetries very difficult. In fact, they have not been observed yet. This motivated several authors to suggest that the observation of these effects, at a higher level than expected, will be an indication of contributions beyond the standard model.

The observation of these effects in the neutral B mesons was experimentally promising because the decay of the b quark to its heavier partner, the top quark, is not possible for kinematic reasons. The suppressed decay width is comparable to the difference in mass of the B_d mesons. Consequently, the mixing over their lifetimes is substantial and has been observed. In addition, the mixing of the states provides another phase that interferes with phases of decay amplitudes and produces oscillations observable in the experiments. These are some of the topics to be covered in this chapter.

16.2 The D^0–$\bar{D}^0$ transition amplitude

The calculation of the mass difference for K_S and K_L mesons in terms of the box diagrams gave a sizable fraction ($\geq$50%) of the observed value. This suggests that similar calculations for heavier mesons may be more accurate. Indeed, estimates of ΔM and $\Delta\Gamma$ for D^0 mesons give small values relative to the decay width of these particles. For this reason, mixing of the neutral D states and CP asymmetries have not yet been observed. Similar calculations for the B mesons give values consistent with the data, because box diagrams are dominated by top quarks in the intermediate states.

As mentioned in Section 14.7, mixing of states depends on the size of the decay width relative to the value of the off-diagonal matrix elements of the effective Hamiltonian. We can estimate both of these terms for D mesons. The decay width in the spectator model is given approximately by

$$\Gamma = \frac{G^2 m_c^5}{192\pi^3}|V_{cs}|^2. \tag{16.1}$$

This should be compared with the following matrix element of the effective Hamiltonian:

$$\begin{aligned}\langle \bar{D}^0|H_{12}|D^0\rangle = &-\frac{G^2}{16\pi^2}M_W^2\left[\xi_s^2 E(x_s) + 2\xi_s\xi_b E(x_s, x_b) + \xi_b^2 E(x_b)\right]\\ &\times\langle \bar{D}^0|\bar{c}\gamma_\mu(1-\gamma_5)u\bar{c}\gamma^\mu(1-\gamma_5)u|D^0\rangle.\end{aligned} \tag{16.2}$$

The factors $\xi_s = V_{us}^* V_{cs}$ and $\xi_b = V_{ub}^* V_{cb}$ are given in terms of Kobayashi–Maskawa matrix elements, which in the Wolfenstein parametrization are of order λ and λ^5, respectively. The $E(x_s)$ and $E(x_b)$ pertain to strange and bottom quarks and are approximated for $m_i^2 \ll M_W^2$ by

$$E(x_i) \approx -x_i. \tag{16.3}$$

The reduced matrix element

$$X_D = \langle \bar{D}|\bar{c}_\alpha\gamma_\mu(1-\gamma_5)u_\alpha\bar{c}_\beta\gamma^\mu(1-\gamma_5)u_\beta|D\rangle, \tag{16.4}$$

with α and β color indices, is similar to the matrix element encountered for K mesons in Section 15.4. An order-of-magnitude estimate is given by the vacuum-insertion approximation, which consists of introducing the vacuum state in all possible ways. For the above case we obtain

$$\begin{aligned}X_D &= \langle \bar{D}|\bar{c}_\alpha\gamma_\mu(1-\gamma_5)u_\alpha|0\rangle\langle 0|\bar{c}_\beta\gamma_\mu(1-\gamma_5)u_\beta|D\rangle\\ &\quad + \langle \bar{D}|\bar{c}_\alpha\gamma_\mu(1-\gamma_5)u_\beta|0\rangle\langle 0|\bar{c}_\beta\gamma_\mu(1-\gamma_5)u_\alpha|D\rangle\\ &= \left(1+\frac{1}{3}\right)|\langle \bar{D}|\bar{c}_\alpha\gamma_\mu(1-\gamma_5)u_\alpha|0\rangle|^2\\ &= \frac{8}{3}F_D^2 M_D.\end{aligned} \tag{16.5}$$

The second equation is obtained by Fierzing the second and the fourth spinors and the factor of $\frac{1}{3}$ by transforming the matrix elements to color singlets (use

Eq. (15.88)). The decay coupling constant is defined as

$$\langle 0|\bar{c}_L \gamma_\mu u_L|D\rangle = \mathrm{i}\frac{F_D\, p_\mu}{\sqrt{2M_D}}, \tag{16.6}$$

with u_L being a normalized left-handed spinor.

It is straightforward to estimate ratios of the mass and width differences to the width. The b-quark exchange graphs are smaller than that for the strange-quark exchange, because of the very small b coupling. Estimates of the strange-quark graph give

$$x = \frac{\Delta M}{\Gamma} \approx O(10^{-4}) \qquad \text{and} \qquad y = \frac{\Delta\Gamma}{2\Gamma} \approx O(10^{-4}), \tag{16.7}$$

with more precise estimates depending on values for the parameters. Alternatively, one may use hadronic intermediate states, but very few amplitudes are known at present and their phases are unknown. The above values of x and y give very small mixing.

Since the lifetime of D mesons is very short, only time-integrated effects can be observed. They are defined in terms of the transition probabilities $|\langle\bar{D}^0|D^0(t)\rangle|^2$ and $|\langle D^0|D^0(t)\rangle|^2$, whose functional forms in terms of widths and masses are similar to those for the K mesons given at the end of Section 15.3:

$$\begin{aligned} r &= \frac{\int_0^\infty |\langle\bar{D}^0|D^0(t)\rangle|^2}{\int_0^\infty |\langle D^0|D^0(t)\rangle|^2} = \left|\frac{q}{p}\right|^2 \frac{\int |f_-(t)|^2\,\mathrm{d}t}{\int |f_+(t)|^2\,\mathrm{d}t} \\ &= \left|\frac{q}{p}\right|^2 \frac{\Delta M^2 + (\Delta\Gamma/2)^2}{2\Gamma^2 + (\Delta M)^2 - (\Delta\Gamma/2)^2} = \left|\frac{q}{p}\right|^2 \frac{x^2+y^2}{2+x^2-y^2}. \end{aligned} \tag{16.8}$$

These equations imply that the expected mixing for neutral D mesons for the values in (16.7) will be of order 10^{-8} and any observable mixing must be attributed to another mechanism beyond the standard model.

Even though the estimates for D^0 are very approximate, the methods we described in this section are general and can be taken over for other mesons. The vacuum-insertion approximation of Eq. (16.5) has already been used for K mesons and we will meet it again in the next section. Vacuum insertion is an approximation that several authors tried to improve. It is hoped that lattice gauge theories will eventually give precise values.

The second result of this section, Eq. (16.8), gives the mixing of short-lived states integrated over long intervals of time. This is a general result that depends on the quantum-mechanical development of a two-state system. In a tagged D^0 beam, the ratio r gives the number of wrong-sign leptons produced in the decays divided by the number of right-sign leptons. The wrong-sign leptons are those which originate from the oscillation of D^0 to $\bar{D}^0$ mesons. We mentioned leptons as an example, but they can be substituted by decays into K mesons of specific strangeness. Such a ratio was in fact used to determine B^0–$\bar{B}^0$ mixing, which we shall explain in Section 16.4.

16.3 Comparison of K^0 and B^0 mesons

It is instructive at the very beginning to compare several properties of the K^0 and B^0 mesons, because these two mesons are quite different. For the K mesons there are two physical states with very different lifetimes:

$$\tau(K_S) = 89.35\,\text{ps} \qquad \text{and} \qquad \tau(K_L) = 51\,700\,\text{ps}. \tag{16.9}$$

This big difference comes about because the mass and width differences of the K mesons are comparable:

$$\Delta M_K = -\frac{1}{2}\,\Delta\Gamma_K = -\frac{1}{2}\Gamma_s. \tag{16.10}$$

For the B mesons the situation is very different. The lifetime of the B mesons is much smaller:

$$\tau(B) = 1.55 \pm 0.06\,\text{ps}. \tag{16.11}$$

In addition $\Gamma_{12} \ll M_{12}$ for B_d mesons, which makes the lifetimes of the two physical states almost identical. For this reason we characterize them by their masses, as heavy and light, and denote them by B_H and B_L, respectively. From the mixing of the two states we know the ratio

$$\frac{\Delta M}{\Gamma} = 0.73 \pm 0.18. \tag{16.12}$$

The mixing of the B states is described by box diagrams analogous to those in Fig. 14.4, with the top quark dominating in the intermediate states. Computation of the diagrams gives the mixing parameter, ε_B, as

$$\frac{q}{p} = \frac{1-\varepsilon_B}{1+\varepsilon_B} \approx \frac{V_{td}}{V_{td}^*} = e^{-2i\beta}, \tag{16.13}$$

with β the phase of the V_{td}^* matrix element; see Eq. (16.27). It follows from this relation that ε_B is mostly imaginary with a small real part. Consequently, the leptonic asymmetry that was useful in K decays is too small, so another method for observing CP violation has had to be discovered.

Among the interesting phenomena are the mixing of B_d states and CP asymmetries which have been observed in decays of these mesons. It is mathematically easier to discuss the CP asymmetry because it deals with the time development for single states $|B^0(t)\rangle$ and $|\bar{B}^0(t)\rangle$. The mixing, on the other hand, observes the correlated development of two states and its calculation is more complicated. Historically, mixing was observed first and the CP asymmetries were observed only very recently. We shall follow the historical development and describe first the mixing of B_d^0 and $\bar{B}_d^0$ states. So far we have mentioned B_d states as typical mesons because experimental results on their decays are available. There are also the B_s mesons, which are very interesting because several of the parameters are different. We postpone a comparison of B_s and B_d states until Section 16.6.

16.4 Mixing in the B_d system

The mixing between B_d and $\bar{B}_d$ mesons was discovered in electron–positron collisions in which a B^0–$\bar{B}^0$ pair is produced. As the produced pair develops in time, the particles oscillate. The time development of each state separately is given by the following equations. A state that starts as $|B^0(t=0)\rangle$ develops according to

$$|B^0(t)\rangle = N\left[f_+(t)|B^0\rangle + \frac{q}{p} f_-(t)|\bar{B}^0\rangle\right]. \tag{16.14}$$

Similarly, a state that starts as $|\bar{B}^0(t=0)\rangle$ develops as

$$|\bar{B}^0(t)\rangle = N\left[\frac{p}{q} f_-(t)|B^0\rangle + f_+(t)|\bar{B}^0\rangle\right], \tag{16.15}$$

with

$$f_\pm(t) = \mathrm{e}^{-\mathrm{i}(M_\mathrm{H}-\mathrm{i}\Gamma_\mathrm{H}/2)t} \pm \mathrm{e}^{-\mathrm{i}(M_\mathrm{L}-\mathrm{i}\Gamma_\mathrm{L}/2)t}. \tag{16.16}$$

and N the normalization factor. To describe the oscillation data properly, we must use quantum-mechanical wave functions for a B^0–$\bar{B}^0$ pair. The pair of B mesons created at the Υ(4S) resonance is a state with odd charge conjugation with the two mesons flying apart from each other with momenta $\vec{k}$ and $-\vec{k}$. The oscillations that set in are highly correlated. The time evolution of the pair is now given by

$$|B^0(t),\,\vec{k}\rangle|\bar{B}^0(t'),\,-\vec{k}\rangle - |\bar{B}^0(t),\,\vec{k}\rangle|B^0(t'),\,-\vec{k}\rangle. \tag{16.17}$$

It is evident that the decays can take place at different times, with the production of leptons through semileptonic decays. For example, we can consider events of the type

$$\begin{array}{llll} \mathrm{e^+e^-} \longrightarrow & \mathrm{B^0} & & +\,\mathrm{\bar{B}^0} \\ & \hookrightarrow \mathrm{Y^+}\ell^-\bar{\nu} & & \hookrightarrow \mathrm{X^-}\ell^+\nu \\ & \quad \text{or } \mathrm{X^-}\ell^+\nu & & \quad \text{or } \mathrm{Y^+}\ell^-\bar{\nu}. \end{array} \tag{16.18}$$

Consequently, events are produced in which the pairs of primary leptons emitted are $\ell^+\ell^+$, $\ell^+\ell^-$, $\ell^-\ell^+$, and $\ell^-\ell^-$. We denote the corresponding rates by l^{++}, l^{+-}, l^{-+}, and l^{--}, respectively. The observation of the parameter

$$R \equiv (l^{++} + l^{--})/(l^{+-} + l^{-+}) \tag{16.19}$$

characterizes particle–antiparticle mixing. We denote by A^{--} the amplitude that one of the mesons with momentum $-k$ decays at time t' into ℓ^- and the other meson decays at time t also to ℓ^-. Using the time dependence of the $|B^0(t)\rangle$ and $|\bar{B}^0(t)\rangle$ states given explicitly in Eqs. (16.14) and (16.15), we obtain for the amplitude A^{--}

$$A^{--}(t', t) = \langle \ell^- Y^+|H|B^0, -k\rangle\langle \ell^- Y^+|H|B^0, +k\rangle(p/q)[f_-(t')f_+(t) - f_+(t')f_-(t)], \tag{16.20}$$

and, in the same notation,

$$A^{-+}(t', t) = \langle \ell^- Y^+|H|B^0, -k\rangle\langle \ell^+ X^-|H|\bar{B}^0, +k\rangle\big[f_-(t')f_-(t) - f_+(t')f_+(t)\big]. \tag{16.21}$$

There are two more equations defining the amplitudes $A^{+-}(t', t)$ and $A^{++}(t', t)$, Paschos and Türke, 1989, p. 218).

For the matrix elements, we introduce the abbreviations

$$\mathcal{M} = \langle \ell^- Y^+|H|B^0, \pm k\rangle, \qquad \bar{\mathcal{M}} = \langle \ell^+ X^-|H|\bar{B}^0, \pm k\rangle,$$

which, according to CPT symmetry, satisfy

$$|\mathcal{M}| = |\bar{\mathcal{M}}| = M. \tag{16.22}$$

It is easy to calculate the rates of decay to each pair of charges by squaring the amplitudes and integrating over the times t and t', separately. After some algebra and a few integrations, the final answer is

$$R = \frac{1}{2}\left(\left|\frac{q}{p}\right|^2 + \left|\frac{p}{q}\right|^2\right)\frac{x^2 + y^2}{2 + x^2 - y^2}, \tag{16.23}$$

with x and y defined for B mesons with $\Delta M = M_{\rm H} - M_{\rm L}$ and $\Delta\Gamma = \Gamma_{\rm H} - \Gamma_{\rm L}$. It is interesting to note that Eq. (16.23) is similar in many respects to Eq. (16.8).

Experiments measured the ratio in e^-e^+ collisions and found

$$R = 0.23 \pm 0.09 \pm 0.03. \tag{16.24}$$

For $|q/p| = 1$ and $y \ll 1$, which will be shown later to be an excellent approximation,

$$\frac{\Delta M}{\Gamma} = 0.73 \pm 0.18. \tag{16.25}$$

It is now a theoretical problem to calculate ΔM and $\Delta\Gamma$ for the B system and compare it with the above values.

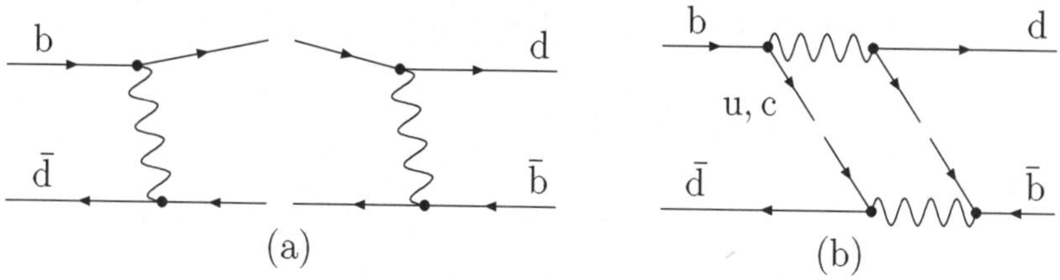

Figure 16.1. Absorptive parts contributing to the B^0–$\bar{B}^0$ width mixing.

The mass difference is given by the box diagrams. Estimates similar to those of the previous sections indicate that the B^0–$\bar{B}^0$ transition amplitude is dominated by the exchange of two top quarks in the box diagram. This is indeed a short-distance contribution. Formula (15.48) translates into

$$M_{12} = -\frac{G^2}{16\pi^2} M_W^2 X_B \xi_t^2 E(x_t)\tilde{\eta}, \tag{16.26}$$

with

$$\xi_t = V_{tb} V_{td}^* = A\lambda^3(1 - \rho - i\eta), \tag{16.27}$$

$$X_B = \langle B_0 | \bar{b}\gamma_\mu(1-\gamma_5)d\bar{b}\gamma^\mu(1-\gamma_5)d|\bar{B}_0\rangle, \tag{16.28}$$

and $\tilde{\eta} \approx 0.87$ is a factor originating from QCD corrections. The reduced matrix element is parametrized in terms of the vacuum-insertion term times a B factor,

$$X_B = \frac{8}{3}|\langle B^0 | \bar{b}\gamma_\mu(1-\gamma_5)d|0\rangle|^2 B_b = \frac{8}{3} F_B^2 M_B B_b. \tag{16.29}$$

The factor $\frac{8}{3}$ originates from the various terms in the product of the currents and rearrangement of the color indices, as was explained in Section 16.2.

Before we consider the magnitude of the mass difference, it will be useful to calculate the width difference computed as the absorptive part of the box diagrams.

For the calculation of the absorptive part we set the intermediate states on the mass shell. This is equivalent to cutting the diagrams in the manner shown in Fig. 16.1 and then integrating over the two-body phase space.

After completion of the integrations and substitution of the reduced matrix element from Eq. (16.29), one obtains (Hagelin, 1981)

$$\Gamma_{12} \simeq -\frac{G^2 m_b^2}{8\pi} F_B^2 M_{B_d} B\left[(\xi_c + \xi_u)^2 - \frac{8}{3}\frac{m_c^2}{m_b^2}(\xi_c^2 + 2\xi_c\xi_u) + O\left(\frac{m_c^4}{m_b^4}\right)\right]. \tag{16.30}$$

This is a relatively simple formula, which can be compared with M_{12}. Neglecting terms of order $(m_c/m_b)^2$, we observe that Γ_{12} has the same phase as M_{12}. This is evident from the unitarity of the Kobayashi–Maskawa matrix, which gives

$$\xi_u + \xi_c = -\xi_t \tag{16.31}$$

and consequently

$$\Gamma_{12} = -\frac{G^2 m_b^2}{8\pi} F_{B_d}^2 M_{B_d} B \xi_t^2. \tag{16.32}$$

Whenever M_{12} and Γ_{12} have the same phase,

$$\frac{q}{p} = \left(\frac{M_{12}^*}{M_{12}}\right)^{\frac{1}{2}} = \frac{V_{td}}{V_{td}^*} = e^{-2i\beta}, \tag{16.33}$$

as was already given in Eq. (16.13). This ratio will be important for the discussion of CP violation.

A second consequence is the magnitude of Γ_{12} relative to M_{12}:

$$\frac{\Gamma_{12}}{M_{12}} \approx 6\pi \left(\frac{m_b}{m_t}\right)^2 \approx 10^{-2}. \tag{16.34}$$

Numerical estimation of the mass difference gives the value $\Delta M \approx 0.73\Gamma$, in agreement with the mixing of the B^0 and $\bar{B}^0$ states.

16.5 Decay rates and CP violation

B-meson decays are frequently described in terms of quark diagrams, which are classified as tree, penguin, or other types of diagrams. The classification is very useful since it specifies how the CP phases appear in amplitudes, that is as couplings of the CKM-matrix elements. The detailed dynamics are not completely understood and we presented several methods for analyzing them in Chapter 14.

In the spectator model, the simplest diagrams for the decay of a $\bar{b}$ involve only one intermediate W^+ boson, as shown in Fig. 16.2.

We denote the decays of the W^+ boson as $u\bar{d}$ or $c\bar{s}$ and we indicate in closed ovals the final hadronic states. Thus the diagrams (a) denote the decays

$$B_d \to D_s^+ D_d^- \qquad \text{or} \qquad B_d \to \pi^+ D_d^-, \tag{16.35}$$

with the couplings $V_{cb}^* V_{cs}$ and $V_{cb}^* V_{ud}$, respectively. Similarly, the decays in diagram (b) are

$$B_d \to J/\psi + K_s \qquad \text{or} \qquad B_d \to D_u^0 \pi^0, \tag{16.36}$$

with the couplings $V_{cb}^* V_{cs}$ and $V_{cb}^* V_{ud}$, respectively.

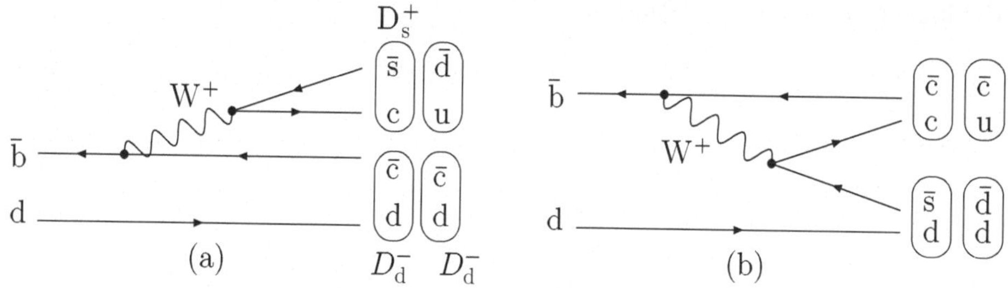

Figure 16.2. Tree-diagram decays of B_d^0.

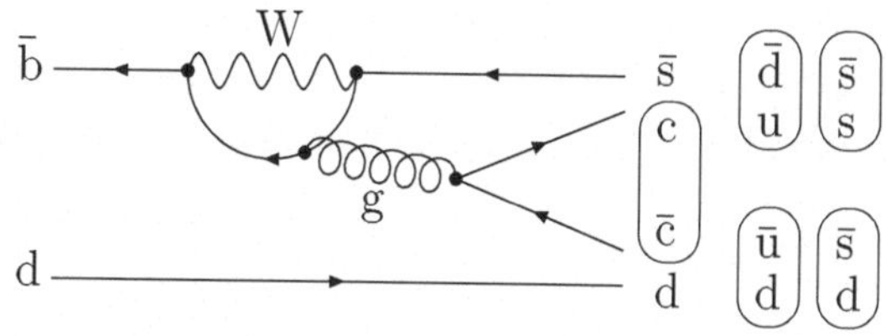

Figure 16.3. A gluonic penguin diagram.

In addition to the tree diagrams, there are penguin diagrams analogous to the ones we discussed in the previous chapter. A typical diagram for gluonic decays is shown in Fig. 16.3.

The final state is determined by the quark assignments. They can produce the following states:

$$B_d \to J/\psi + K_s, \quad B_d \to \phi K_s, \tag{16.37}$$

$$B_d \to \pi^+\pi^-, \ \pi^0\pi^0. \tag{16.38}$$

We see that the $J/\psi + K_s$ decay can originate from tree and penguin diagrams whose dynamics are very different. The dominant penguin diagrams are those with charm and top quarks in the intermediate states (see Problem 4) whose couplings are

$$V_{cb}^* V_{cs} = A\lambda^2 + O(\lambda^4), \tag{16.39}$$

and

$$V_{tb}^* V_{ts} = -A\lambda^2, \tag{16.40}$$

respectively. This CKM coupling can be extracted as a multiplicative factor. The property is unique to this decay channel and makes the predictions very reliable. For this reason the decay has been named the gold-plated decay channel. The mode has several advantages.

(i) The prediction for the CP asymmetry is reliably expressed in terms of CKM parameters and is large.

(ii) It has a measurable branching ratio

$$\mathrm{Br}(\mathrm{B} \to \mathrm{J}/\psi \mathrm{K}^0) = (8.7 \pm 0.5) \times 10^{-4}, \tag{16.41}$$

with the decays $\mathrm{J}/\psi \to \pi^+\pi^-$ and $\mathrm{K}^0 \to \pi^+\pi^-$ producing hadrons, which are easily detectable.

(iii) The final state $\mathrm{J}/\psi + \mathrm{K_s}$ is a CP eigenstate.

For these reasons the decay of B^0 mesons to $\mathrm{J}/\psi + \mathrm{K_s}$ is an attractive mode and we describe the decay amplitudes in Problem 4.

Fortunately there is also an experimental method for detecting CP-violating effects (Carter and Sanda, 1981; Bigi and Sanda, 1981). A $\mathrm{B}^0\bar{\mathrm{B}}^0$ pair is produced in electron–positron colliders and proceeds to decay. We select a decay mode for B^0 and $\bar{\mathrm{B}}^0$ to a common final state $|f\rangle$, which is an eigenstate of CP. Let

$$A_{\mathrm{f}} = \langle f|H|B^0\rangle \qquad \text{and} \qquad \bar{A}_{\mathrm{f}} = \langle f|H|\bar{B}^0\rangle$$

be the decay amplitudes. Since the beams of particles created in the collider are a mixture of B^0 and $\bar{\mathrm{B}}^0$ and since their lifetimes are almost identical, we cannot a priori tell whether $|f\rangle$ arose from the decay of B^0 or $\bar{\mathrm{B}}^0$. Thus we need independent information on the flavor of the decaying neutral B^0 meson. This can be achieved by observing a semileptonic decay on one side and the decay to $|f\rangle$ on the other side. In this way we know whether the decay to $|f\rangle$ originates from a B^0 or $\bar{\mathrm{B}}^0$. The experimental groups measure these decays as a function of time. A small asymmetry in the time evolution of the two decays is evidence for CP violation.

We describe the time evolution of particle and antiparticle decays in detail. The decay amplitudes as a function of time are

$$\langle f|B^0(t)\rangle = \frac{1}{2}\left[f_+(t)\,A_{\mathrm{f}} + \frac{q}{p}f_-(t)\bar{A}_{\mathrm{f}}\right], \tag{16.42}$$

$$\langle f|\bar{B}^0(t)\rangle = \frac{1}{2}\left[f_-(t)\,A_{\mathrm{f}} + \frac{q}{p}f_+(t)\bar{A}_{\mathrm{f}}\right]. \tag{16.43}$$

The amplitude A_{f} is the expectation value of the quark operators between $|B\rangle$ and $\langle f|$ states. It is computed with the help of quark diagrams described at the beginning of this section. They have a weak phase coming from the CKM-matrix elements and perhaps a phase of strong origin from final-state interactions. We shall denote the ratio

$$\rho = \frac{\bar{A}_{\mathrm{f}}}{A_{\mathrm{f}}}. \tag{16.44}$$

In the case of the B^0 mesons the factors $f_\pm(t)$ simplify, because

$$\Gamma_{\mathrm{H}} = \Gamma_{\mathrm{L}} = \Gamma$$

and

$$f_{\pm}(t) = \mathrm{e}^{-\frac{1}{2}\Gamma t}\left(\mathrm{e}^{-\mathrm{i}M_{\mathrm{H}}t} \pm \mathrm{e}^{-\mathrm{i}M_{\mathrm{L}}t}\right). \tag{16.45}$$

The rate for detecting a decay to $|f\rangle$ at a time t after the production of a B^0 is proportional to

$$\Gamma(t) \approx \frac{1}{2}\mathrm{e}^{-\Gamma t}[1 - S_{\mathrm{f}}\sin(\Delta m\, t) + C_{\mathrm{f}}\cos(\Delta m\, t)] \tag{16.46}$$

and, for the detection of a decay from an initial $\bar{\mathrm{B}}^0$,

$$\bar{\Gamma}(t) \approx \frac{1}{2}\mathrm{e}^{-\Gamma t}[1 + S_{\mathrm{f}}\sin(\Delta m\, t) - C_{\mathrm{f}}\cos(\Delta m\, t)], \tag{16.47}$$

with

$$S_{\mathrm{f}} = \frac{2\,\mathrm{Im}\left(\frac{q}{p}\rho\right)}{1 + \left|\frac{q}{p}\rho\right|^2} \quad \text{and} \quad C_{\mathrm{f}} = \frac{1 - \left|\frac{q}{p}\rho\right|^2}{1 + \left|\frac{q}{p}\rho\right|^2}. \tag{16.48}$$

In both of these equations there is a time-oscillation superimposed on the exponential decay. The formulas are very general and can be applied to several decays. We consider some decays in detail.

(a) The gold-plated channel. For the decay $\mathrm{B} \to \mathrm{J}/\psi + \mathrm{K_s}$ there are tree and penguin diagrams. As we discussed in this section, all dominant diagrams have zero CKM phase (see Eqs. (16.36) and (16.39)). In addition, q/p is given by Eq. (16.13) and we obtain

$$S_{\mathrm{f}} = -\sin(2\beta), \qquad C_{\mathrm{f}} = 0$$

and

$$\Gamma(t) \approx \mathrm{e}^{-\Gamma t}[1 + \sin(2\beta)\sin(\Delta m\, t)]. \tag{16.49}$$

Thus the time evolution of the $\mathrm{B}_{\mathrm{d}}^0$ state and its conjugate particle to a common final state is an efficient method for identifying CP parameters. The asymmetry

$$\alpha = \frac{\Gamma(t) - \bar{\Gamma}(t)}{\Gamma(t) + \bar{\Gamma}(t)} = \sin(2\beta)\sin(\Delta m\, t) \tag{16.50}$$

has a sinusoidal dependence on time. This decay has been measured in the BaBar and Belle experiments, giving the average value

$$\sin(2\beta) = 0.73 \pm 0.03. \tag{16.51}$$

The angle β extracted from the asymmetry is one of the angles in the unitarity triangle. It enters the calculation through the element V_{td} given in Eq. (9.53).

The unitarity triangle is constrained by other measurements as well; to be precise, by the magnitude of V_{ub}, the parameter ε_k, and the mixing of $\mathrm{B_d}$ and $\bar{\mathrm{B}}_{\mathrm{d}}$ states. Each of these quantities determines a region and their intersection defines the apex of the triangle. This figure has become very popular and is featured in Fig. 9.2 and in many articles (Branco *et al.*, 1999; Kleinknecht, 2003; Brower and Faccini, 2003). The value of β in Eq. (16.51) has several solutions and one of them coincides with a direction that goes through the apex of the triangle. The significance of the $\sin(2\beta)$ measurement is that for the first time a large CP asymmetry has been observed, proving that CP is not an approximate symmetry of nature.

(b) Decays to $\pi^+\pi^-$ and $\pi^0\pi^0$. The $B \to \pi\pi$ decays can be analyzed in a similar manner to the $K \to \pi\pi$ decays. For the sake of brevity we shall use a similar notation; however, the numerical values for the quantities (amplitudes, phases, etc.) in B decays are different. The amplitude for $B^0 \to \pi^+\pi^-$ is written as

$$A_f(\pi^+\pi^-) = \sqrt{\frac{2}{3}}|A_0|e^{i(\delta_0-\theta_0)} - \frac{2}{\sqrt{3}}|A_2|e^{i(\delta_2-\theta_2)} \tag{16.52}$$

and

$$\bar{A}_f(\pi^+\pi^-) = \sqrt{\frac{2}{3}}|A_0|e^{i(\delta_0+\theta_0)} + \frac{1}{\sqrt{3}}|A_2|e^{i(\delta_2+\theta_0)}. \tag{16.53}$$

The phases δ_0 and δ_2 come from strong interactions of the final states, but at this high energy they cannot be related to $\pi\pi$ phase shifts. The phases θ_0 and θ_2 come from weak interactions, which originate from CKM couplings. The tree diagram contributes to both the $I = 0$ and the $I = 2$ amplitude and the penguin diagram only to the $I = 0$ amplitude.

The attempt to determine the isospin amplitudes by comparing decays of charged and neutral B mesons has met with limited success. This approach is analogous to that for the $K \to \pi\pi$ decays, in which, after the isospin analysis of the amplitudes, we had to return to the effective QCD Hamiltonian.

There are also analyses in terms of Feynman diagrams. We have already noted that this decay mode receives contributions from tree and penguin diagrams. In the Wolfenstein parametrization all diagrams are of order λ^3. The tree diagram has the phase $e^{-i\gamma}$. The penguin diagrams have intermediate states with up, charm, and top quarks, each with a different phase. The relevant parameter for this decay has the general form

$$\frac{q}{p}\frac{A_f(\pi^+\pi^-)}{\bar{A}_f(\pi^+\pi^-)} = e^{-2i\beta}\frac{e^{-i\gamma} + he^{i\theta}}{e^{+i\gamma} + he^{i\theta}}, \tag{16.54}$$

with h being the ratio of a term from the penguin diagram to the remaining contributions and θ a phase of strong origin. The interference of mixing with the decay amplitude gives

$$S_f(\pi^+\pi^-) \neq 0 \quad \text{and} \quad C_f(\pi^+\pi^-) \neq 0. \tag{16.55}$$

Consequently, the presence of both $\cos(\Delta m\, t)$ and $\sin(\Delta m\, t)$ terms with coefficients different from zero is an indication of CP violation in the B amplitudes. For very small values of h, this mode fixes

$$S_f = \sin(2\beta + 2\gamma),$$

which, through the triangle relation $\beta + \gamma = \pi - \alpha$, can be replaced by α. Thus this decay mode can determine $\beta + \gamma$ and, indirectly, the angle α.

(c) Decays to other channels. In the previous discussion we demonstrated that β is accurately determined and γ, or alternatively α, can be extracted from the B $\to \pi\pi$ decay, provided that the hadronic matrix elements are better understood. There are extensive efforts to analyze and understand other decay modes. For instance (Fleischer, 2003),

$$\mathrm{B} \to \phi + \mathrm{K_s}$$

has only penguin contributions. An analysis similar to that of the $\mathrm{J}/\psi\mathrm{K_s}$ mode reveals many similarities. In the absence of new physics, the asymmetries for charmonium and ϕ-meson decays should be equal.

This heralds a new era, in which the theory is expected to be scrutinized through a variety of other $\mathrm{B_d}$ and $\mathrm{B_s}$ decay asymmetries. Impressive progress is being made in the search for asymmetries in the aforementioned modes and in addition $\mathrm{B}^{0\pm} \to \mathrm{K}\pi$ and $\mathrm{B}^{\pm} \to \mathrm{DK}^{\pm}$. Current measurements are approaching the stage of cross-checking the theory or even discovering physical phenomena beyond the standard model.

16.6 Mass and lifetime differences for $\mathrm{B_s}$ mesons

The analysis of the previous sections can be extended in a straightforward way to the $\mathrm{B_s}$ mesons, which present a very interesting mesonic system. Their masses and lifetimes are very similar to those of the $\mathrm{B_d}$ mesons, but their weak interactions are different because the CKM couplings are now much bigger. A consequence is the larger mixing in these states and, it is hoped, a lifetime difference between two physical particles that may be measurable. The reduced matrix element is given now as

$$X_{\mathrm{B_s}} = \frac{8}{3}|\langle B_{\mathrm{s}}^0|\,\bar{b}\gamma_\mu(1-\gamma_5)s\,|0\rangle|^2 B_{\mathrm{s}} \tag{16.56}$$

and is expected to have a numerical value close to $X_{\mathrm{B_d}}$ because the wave functions and the general structure of the bound state are expected to be similar. The coupling which appears in the box diagrams is

$$\xi_{\mathrm{t}}' = V_{\mathrm{tb}} V_{\mathrm{ts}}^* = -A\lambda^2, \tag{16.57}$$

which is much larger than the coupling in Eq. (16.27). The mass difference is given by a formula analogous to Eq. (16.26) and, on taking the ratio, one obtains

$$\frac{\Delta M_{\mathrm{s}}}{\Delta M_{\mathrm{d}}} = \frac{F_{\mathrm{s}}^2}{F_{\mathrm{d}}^2}\frac{M_{\mathrm{s}}}{M_{\mathrm{d}}}\left|\frac{V_{\mathrm{ts}}}{V_{\mathrm{td}}}\right|^2. \tag{16.58}$$

The ratio of the hadronic matrix parameters is unity in the SU(3) limit or the heavy-quark limit. A small deviation from unity may still show up, but this will be of order

10%–20%. The large change comes from the couplings

$$\left|\frac{V_{\mathrm{ts}}}{V_{\mathrm{td}}}\right|^2 = \frac{1}{\lambda^2\left[(1-\rho)^2+\eta^2\right]} = 20\text{–}30. \tag{16.59}$$

At present there is an experimental bound, $\Delta M_{\mathrm{s}}/\Delta M_{\mathrm{d}} > 21.2$, coming from the mixing of $\mathrm{B}_{\mathrm{s}}^0$ and $\bar{\mathrm{B}}_{\mathrm{s}}^0$-particles, which is close to 100%. Serious experimental efforts are being devoted to searching for a precise value of the mixing, which will also determine the ratio $x_{\mathrm{s}} = \Delta M/\Gamma$. Since the mixing approaches 100%, a precise measurement is required in order to extract a value for the mass difference (see Eq. (16.23)).

The estimate which makes x_{s} large also increases the width difference of the two states. The two B_{s} states can mix to form two distinct eigenstates. To a first approximation,

$$\frac{\Delta\Gamma_{\mathrm{s}}}{\Delta M_{\mathrm{s}}} = \frac{3}{2}\pi\left(\frac{m_{\mathrm{b}}}{m_{\mathrm{t}}}\right)^2 \approx 3.7\times 10^{-3} \tag{16.60}$$

has been calculated from equations analogous to Eqs. (16.26) and (16.32). The width difference can be rescaled to give

$$\left(\frac{\Delta\Gamma}{\Gamma}\right)_{\mathrm{s}} = \frac{3}{2}\pi\left(\frac{m_{\mathrm{b}}}{m_{\mathrm{t}}}\right)^2\left(\frac{\Delta M_{\mathrm{s}}}{\Delta M_{\mathrm{d}}}\right)\left(\frac{\Delta M_{\mathrm{d}}}{\Gamma}\right)_{\mathrm{d}}\left(\frac{\Gamma_{\mathrm{d}}}{\Gamma_{\mathrm{s}}}\right). \tag{16.61}$$

The ratio of the mass difference was estimated in this section, and assuming $\Gamma_{\mathrm{d}} \approx \Gamma_{\mathrm{s}}$ leads to a width difference that is experimentally interesting. Its value is in the range 7%–14%. Alternative estimates that saturate Γ_{12} with physical intermediate states or form other ratios also give encouraging results. There is a good chance that the lifetime difference $\tau_{\mathrm{H}} - \tau_{\mathrm{L}}$ is 10% of the lifetime of B_{d} mesons. Such a large value would be measurable in the decay

$$\mathrm{B}_{\mathrm{s}} \to \mathrm{J}/\psi + \phi.$$

Problems for Chapter 16

1. Carry out the integrals $\int_0^\infty |f_\pm(t)|^2\,\mathrm{d}t$ and show that the last term in Eq. (16.8) follows. Hint: it is easier to integrate exponentials and then take real parts.
2. Write the matrix element for the mixing of the $\mathrm{B}_{\mathrm{d}}^0 \to \bar{\mathrm{B}}_{\mathrm{d}}^0$ system, especially the CKM couplings of the box diagram with the top quark in the intermediate states. Then prove the ratio

$$\frac{q}{p} = \left(\frac{H_{21}}{H_{12}}\right)^{\frac{1}{2}} \approx \frac{V_{\mathrm{td}}}{V_{\mathrm{ts}}^*}.$$

3. Calculate the amplitude for the diagram in Fig. 16.1(a) with charm and up quarks in the intermediate states. Then integrate over the two-body phase space to obtain some of the terms in Eq. (16.30).

4. The decay $B_d \to J/\psi + K_s$ has tree and penguin diagrams. We denote the tree diagram by

$$T = V_{cb}^* V_{cs} g$$

and the penguin diagram by

$$P = V_{tb}^* V_{ts} f(m_t) + V_{cb}^* V_{cs}\, f(m_c) + V_{ub}^* V_{us} f(m_u),$$

where $f(m_q)$ are functions from the calculation of the penguin diagram with m_q the mass of the q quark in the loop. A naive order-of-magnitude estimate of the tree and loop diagrams gives (Gronau, 1992)

$$\frac{f(m_t)}{g} \sim \frac{\alpha_s(m_b)}{6\pi} \ln\left(\frac{m_t}{m_b}\right) \sim 0.04.$$

Others obtained larger values by computing the penguins with the effective Hamiltonian (Kramer and Palmer, 1995).

(i) Use the unitarity of the CKM matrix to eliminate $V_{tb}^* V_{ts}$ and rewrite P in terms of the other two CKM factors.

(ii) Show that the ratio of the two CKM factors is

$$\left|\frac{V_{ub}^* V_{us}}{V_{cb}^* V_{cs}}\right| \sim \lambda^2 + O(\lambda^4).$$

Prove that the total amplitude

$$\langle J/\psi K_s | H | B \rangle = V_{cb}^* V_{cs}[g + (f(m_c) - f(m_t))],$$

which to order λ^2 is real.

(iii) Use this form of the matrix element to calculate the asymmetry in Eq. (16.50).

5. Analyze the $B \to \phi + K_s$ decay along the lines of Problem 4 and derive Eq. (16.54) with an explicit expression for the function h.

References

Bigi, I. I., and Sanda, A. I. (1981), *Nucl. Phys.* **B193**, 85

Branco, G., Lavoura, L., and Silva, J. P. (1999), *CP Violation* (Oxford, Oxford University Press), Section 18.5

Brower, T. E., and Faccini, R. (2003), *Ann. Rev. Nucl. Particle Sci.* **53**, 353

Carter, A. B., and Sanda, A. I. (1981), *Phys. Rev.* **D23**, 1527

Fleischer, R. (2003), hep-ph/0310313

Gronau, M. (1992), *Phys. Lett.* **B300**, 163; hep-ph/9209279

Hagelin, J. (1981), *Nucl. Phys.* **B193**, 123

Kleinknecht, K. (2003), *Uncovering CP Violation (Experimental Classification in the Neutral K Meson and B Meson Systems)* (Springer Tracts in Modern Physics) (Berlin, Springer-Verlag), pp. 130–135

Kramer, G., and Palmer, W. (1995), *Phys. Rev.* **D52**, 6411

Paschos, E. A. and Türke, U. (1989), *Phys. Rep.* **178**, 145–260

Select bibliography

The books recommended in Chapter 15 discuss many properties of D and B mesons.

17

Higgs particles

17.1 Higgs-boson couplings

The electroweak theory depends crucially on the Higgs mechanism. Many aspects of the theory have been tested in experiments to a high degree of accuracy, especially properties of the gauge bosons and their couplings to fermions. However, the Higgs particles have not been discovered yet. In Chapter 7 we discussed the simplest and, perhaps, the most natural way of breaking the SU(2) symmetry, namely by introducing a doublet of scalar particles. Among them three fields were eliminated, becoming the longitudinal degrees of freedom for the gauge bosons. The remaining neutral particle is physical and should be observed. In the same chapter it was shown that giving the Higgs field a vacuum expectation value generates masses for $W^{\pm}$ and Z^0, whose ratio is related to the weak mixing angle. This ratio has been confirmed experimentally, which provides strong support for the underlying SU(2) symmetry.

The Higgs doublet contains two complex fields:

$$\begin{pmatrix} \phi^+ \\ \phi^0 \end{pmatrix}. \tag{17.1}$$

The breaking of the symmetry introduces the vacuum expectation value

$$\langle \phi^0 \rangle = \frac{v}{\sqrt{2}}. \tag{17.2}$$

The scalar Lagrangian introduced in Chapter 7 has two parameters, μ and λ, which are related to the value of the field at the minimum:

$$v = \frac{\mu}{\sqrt{\lambda}} = \left(\sqrt{2}G\right)^{-\frac{1}{2}}.$$

The expansion of the field around the minimum of the potential

$$\phi' = \frac{1}{\sqrt{2}} \begin{pmatrix} 0 \\ v + H \end{pmatrix} \tag{17.3}$$

introduces the physical field H as a fluctuation around the minimum. The numerical value of v was determined in Eq. (8.20), leaving the mass of the scalar as the only undetermined parameter of the Higgs sector,

$$\frac{M_{\rm H}^2}{2} = -\frac{\mu^2}{2} + \frac{3}{2}\lambda v^2 = \mu^2. \tag{17.4}$$

Consequently, all couplings of the Higgs particles are rewritten in terms of $M_{\rm H}$ and other coupling constants already determined in previous chapters. After a short study of the couplings, one finds

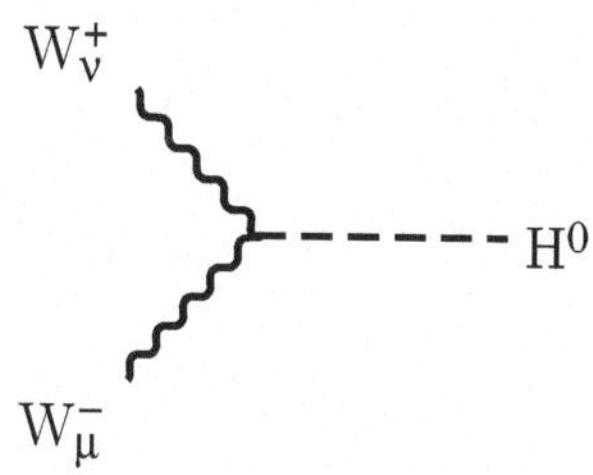

$ig M_{\rm W} g_{\mu\nu}$

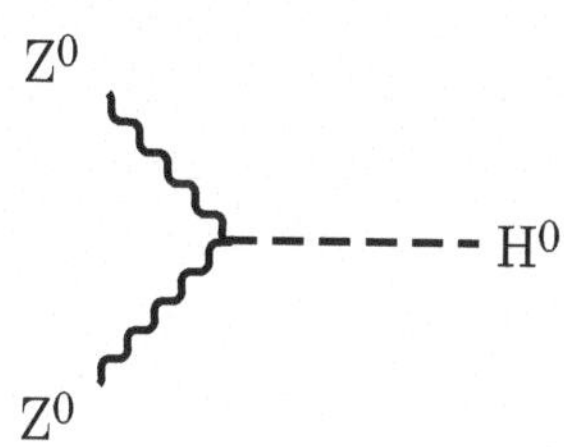

$\frac{ig M_{\rm Z}}{\cos\theta_{\rm W}} g_{\mu\nu}$

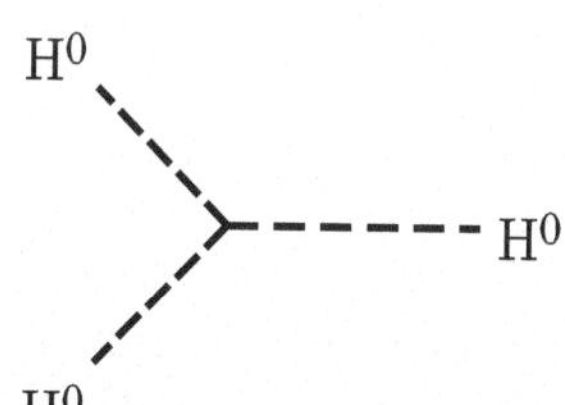

$-i3!\lambda v$

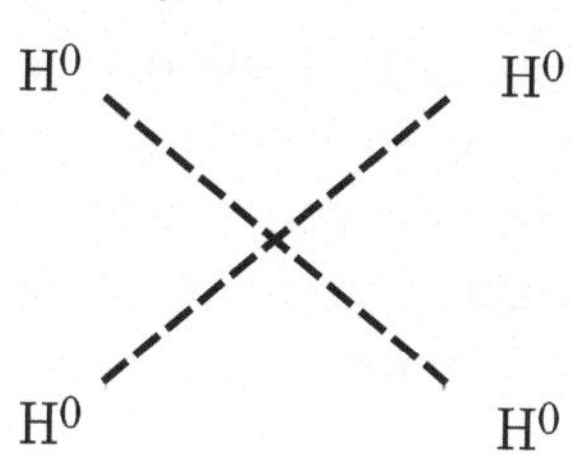

$-i3!\lambda$

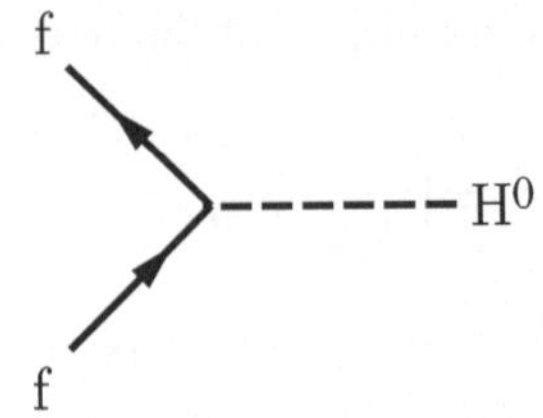

$$\frac{-\mathrm{i}m_{\mathrm{f}}}{v}$$

with

$$v = \frac{\mu}{\sqrt{\lambda}} = (\sqrt{2}G)^{-1/2} = 246\,\mathrm{GeV}.$$

With so many couplings depending on a single parameter, we can proceed to calculate the decays of Higgses to other particles as well as their production cross sections. In this way we find processes in which the Higgs particles must be produced, provided that they are not very heavy, or they should appear in the decay of ordinary or new heavy particles. All these approaches have been pursued actively, but no Higgs particles have been discovered yet. Instead we have bounds on their masses and the production rates. We shall cover several of these topics in this chapter.

A very important role in all such studies is played by the Higgs mass. There are efforts to limit the mass by appealing to unitarity or demanding that the electroweak theory remains renormalizable. An alternative direction of investigation studies the effects introduced by radiative corrections to several processes. Higgs particles, together with quarks and gauge bosons, appear as intermediate states of Feynman diagrams. Many experiments have reached a high level of accuracy so that radiative corrections must be included in order to bring agreement between theory and experiment. They provide a crucial test of the theory and impose constraints on the Higgs mass. We study several of them in the next section.

17.2 Precision tests of the theory

The electroweak theory introduces several parameters that are undetermined. Among them are the mass of the Higgs and the masses and mixing angles of the fermions. The gauge couplings and the masses of the gauge bosons enter in many reactions, where they are determined precisely to have the same value. As two examples we mention the weak mixing angle $\sin^2\theta_{\mathrm{W}}$ and the Fermi coupling constant. They have been measured in various reactions so precisely that higher-order corrections are necessary. The loop corrections include quantum corrections of the theory testing it to a higher degree of accuracy. In this section we review the precise measurements for several parameters and point out reactions where discrepancies may appear on the horizon. At present the agreement is so good that we can consider the results from such tests a great success of the theory.

In order to make predictions, we use three quantities as input parameters. They are defined in various processes where they can be determined precisely, including radiative corrections, and then they are used to predict other processes.

(i) The first is the fine-structure constant α measured in experiments that involve the Josephson junction or the anomalous magnetic moment of the electron. It receives corrections from strong interactions and varies with the momentum at which it is measured in a way similar to the strong coupling constant described in Section 11.2.
(ii) The second parameter is the Fermi coupling constant determined from the muon lifetime. The corrections include electromagnetic and weak effects.
(iii) As a third parameter, we can select one of the masses M_W or M_Z. It is customary to choose M_Z determined from the Z lineshape, because it has been measured more accurately.

With these three parameters we can predict other quantities. For instance,

$$\sin^2\theta_\mathrm{W}\,\cos^2\theta_\mathrm{W} = \frac{\pi\alpha}{\sqrt{2}G_\mathrm{F}}\frac{1}{M_\mathrm{Z}^2} \tag{17.5}$$

and $M_\mathrm{W} = M_\mathrm{Z}\cos\theta_\mathrm{W}$, which is also rewritten as

$$\rho = \frac{M_\mathrm{W}^2}{M_\mathrm{Z}^2\cos^2\theta_\mathrm{W}} = 1. \tag{17.6}$$

The ρ parameter determines the strength of neutral currents relative to charge currents and to lowest order (tree level) it has the value unity. It is modified by corrections, as we will mention below. Measuring quantities and carrying out precision tests of the theory is a research field in itself. I decided to present here the general features that enter such calculations and mention a few comparisons, especially those for which there are small disagreements. The reader who wishes to specialize in this field can consult the corresponding section in the particle data group (Erler and Langacker, 2004) or books devoted to this topic (Bardin and Passarino, 1999).

We have seen in previous chapters that one-loop radiative corrections involve integrals of the form

$$\int \mathrm{d}^4k\,\frac{1}{(k^2+m^2+\mathrm{i}\varepsilon)^\alpha} \tag{17.7}$$

(see, for instance, Sections 14.7 and (15.7)). When $\alpha \leq 2$ the integral diverges. In a renormalizable theory the infinities are absorbed as corrections to masses and coupling constants. Since they are arbitrary, it appears, at first glance, that the corrections are unobservable. An exception to this rule occurs when masses and

Figure 17.1. Self-energies for gauge bosons.

coupling are not free but are related to each other. As a first example, we mention the equality of the vector couplings occurring in muon decay and β-decay. They are used to determine a precise value for V_{ud} as given in Section 9.3.1.

Another accurate example is provided by the ρ parameter in Eq. (17.6). It receives corrections from the self-energies shown in Fig. 17.1, where the intermediate solid lines represent quarks and the dotted lines Higgs bosons. It is sensitive to the masses of particles in the intermediate states and provided a benchmark for restricting the masses of these particles. The above-mentioned corrections give rise to deviations from unity given by the equation

$$\rho = 1 + \frac{G_F}{8\pi^2}\left[m_1^2 + m_2^2 - \frac{2m_1^2 m_2^2}{m_2^2 - m_1^2}\ln\left(\frac{m_2^2}{m_1^2}\right) - \frac{3M_W \sin^2\theta_W}{\cos^2\theta_W}\ln\left(\frac{M_H^2}{M_W^2}\right)\right], \tag{17.8}$$

with m_2 and m_1 the masses of the top and bottom quarks, respectively, and M_H the mass of the Higgs boson. Other quark pairs also contribute, but the magnitude of their correction is smaller.

With the above inputs, $\sin^2\theta_W$ and M_W can be calculated when values for m_t^2 and M_H are given. Such arguments were used to constrain the top quark's mass before its discovery. Experimental results from the Large Electron–Positron Collider (LEP) combined with loop contributions restricted the mass m_t to within the range 160–180 GeV where the top quark was discovered. The standard-model prediction now is

$$m_t(m_t) = 174.3 \pm 3.4\,\text{GeV}. \tag{17.9}$$

Furthermore, the precision electroweak data accumulated at the LEP, SLAC, and the Tevatron strongly support the standard model with a weakly coupled Higgs boson. As described above, the Higgs boson contributes to the $W^{\pm}$ and Z vacuum polarization through loop effects. The result of a global fit is shown in Fig. 17.2 and yields the value (Eidelman, 2004)

$$M_H = 126^{+73}_{-48}\,\text{GeV}, \tag{17.10}$$

which is consistent with the lower bound of 114.4 GeV established in direct searches. These values provide a benchmark for designing experiments aiming at the discovery of the Higgs particles.

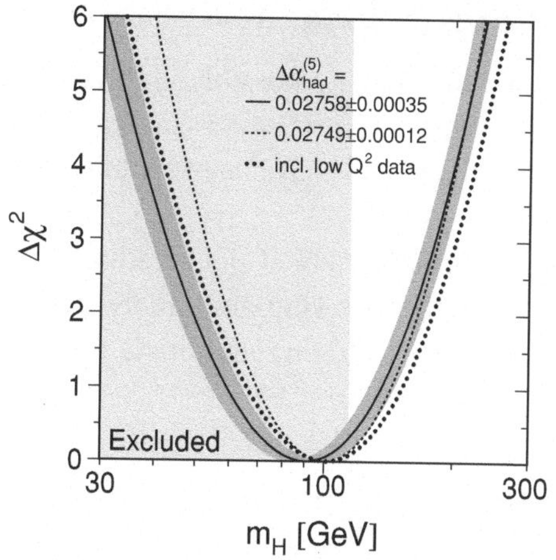

Figure 17.2. The result of the global fit presented as $\Delta\chi^2 = \chi^2 - \chi^2_{\text{min}}$ versus M_{H}.

Before leaving this section, it is worth mentioning two cases in which, if we look at the results in detail, they are either not satisfactory or could indicate small disagreements with the predictions of the standard model. A neutrino experiment at Fermilab completed a precise measurement of $\sin^2\theta_{\text{W}}$ using the R ratio of Eq. (12.17) and other ratios. A 3σ deviation from the standard-model prediction was reported. This conclusion depends on several theoretical parameters entering the analysis that are now under active investigation (Gluck *et al.*, 2005).

A further discrepancy appears in the determination of the effective weak mixing angle. The adjective "effective" indicates that a specific renormalization scheme has been introduced for the higher-order corrections. The two most precise measurements of $\sin^2\theta_{\text{eff}}$ from SLAC from polarized-electron asymmetry A_{LR} and from the forward–backward asymmetry A^{b}_{FB} in the production of b–$\bar{\text{b}}$ pairs differ by $\sim 3\sigma$. In fact, the result from A_{LR} is in good agreement with the leptonic asymmetries measured at the LEP, whereas all other hadronic asymmetries are better compatible with the A^{b}_{FB}.

Besides the efforts for discovering the Higgs bosons, it is also important to confirm or eliminate these discrepancies. There are theoretical corrections that have been pointed out and are being actively discussed. Their implications are so profound that one should continue the investigations. The absence of Higgs particles in the mass range discussed in this section and/or the persistence of the discrepancies will be an indication of new physics. A promising candidate for new physics is supersymmetry, whose presence is capable of modifying the predictions described in this section.

17.3 Bounds on masses from general principles

There are many new reactions that are possible in the electroweak theory. At tree level the amplitudes involve a few partial waves and must satisfy unitarity bounds. For elastic WW scattering the bounds have been exploited effectively to restrict the mass of the Higgs boson.

At very high energies the longitudinal polarization of the W bosons is $\varepsilon_{\mathrm{L}}^{\mu} \approx p^{\mu}/M_{\mathrm{W}}$. It produces the fastest growing amplitudes as functions of energy and involves one partial wave. The scattering amplitude can be expanded in terms of partial waves:

$$M(s,t) = \frac{1}{k}\sum_{\ell=0}^{\infty}(2\ell+1)a_{\ell}(s)P_{\ell}(\cos\theta). \tag{17.11}$$

The fact that each partial wave must satisfy a unitarity bound leads to the condition

$$|a_0| \leq \frac{1}{2} \tag{17.12}$$

and supplies an upper bound for the mass of the Higgs boson (Lee *et al.*, 1977),

$$M_{\mathrm{H}}^2 < \frac{4\pi\sqrt{2}}{G_{\mathrm{F}}} \approx (1.2\,\mathrm{TeV})^2. \tag{17.13}$$

This is the simplest bound described also in Problem 2. It can be improved (Lee *et al.*, 1977) by considering the coupled channels $\mathrm{W_L W_L}$, $\mathrm{Z_L Z_L}$, $\mathrm{Z_L H}$, and HH, with the subscript L indicating longitudinal polarization, then demanding that the eigenvalues of these coupled channels satisfy the unitarity conditions. The improved bound is smaller by a factor of $1/\sqrt{3}$, i.e.

$$M_{\mathrm{H}} \leq 700\,\mathrm{GeV}.$$

New bounds are obtained by requiring stability of the vacuum. We saw in Chapter 5 that the Higgs potential must have the specific form of Fig. 5.1 in order to be able to expand the field around a minimum of the potential. This property must be preserved even when radiative corrections are included. The coupling constant λ receives corrections from loop diagrams and it becomes a "running" coupling depending on the scale Λ at which it is evaluated. The change of λ is given by a differential equation that depends on the Yukawa coupling (quark loops) and the gauge coupling (gauge loops). The corrected potential has the form

$$V_{\mathrm{eff}} = -\mu^2 H^2 + \lambda(\Lambda)H^4 \tag{17.14}$$

and its minimum defines again a vacuum expectation value: $\langle\phi\rangle = v/\sqrt{2}$. If the corrections are large, it is possible to obtain a negative λ and consequently an unstable ground state. On the other hand, for λ large and positive there is one trivial minimum at the origin. Since $V(0) = 0$, a non-trivial minimum exists provided that $V(v/\sqrt{2}) < 0$. This condition gives the lower bound (Linde, 1976)

$$M_{\mathrm{H}} \gtrsim 50\,\mathrm{GeV} \quad \text{with} \quad \Lambda = 1\,\mathrm{TeV}.$$

Experimental searches for Higgses have gone much above this value. The best value available comes from the reaction in Eq. (17.15) to be discussed later on.

17.4 Decays

At the beginning of this chapter we listed several couplings that depend on a single parameter M_{H}. A characteristic property in models with one Higgs doublet is that Higgs particles couple strongly to the heaviest particle. For instance, the couplings to fermions prefer the top quark and the couplings to W and Z bosons are proportional to their masses. With so many couplings available, it is straightforward to compute decay widths and production rates. We list in Table 17.1 decay widths for several channels.

The decays have special properties worth mentioning. The lifetime of the Higgs particle is very short and, being neutral, it leaves no visible track. Its decay products, such as $\mathrm{b\bar{b}}$ or $\tau\tau$ pairs, produce detectable tracks. They are used as signatures for discovering the Higgs boson.

The decays to vector mesons also have several interesting properties. The formulas for decay widths into the weak vector bosons Z and W look quite similar, except for an additional factor of $\frac{1}{2}$, accounting for the symmetric final state in the case of two identical Z bosons. In the limit $m_{\mathrm{Z}}, m_{\mathrm{W}} \ll m_{\mathrm{H}}$ the width of the ZZ pairs is half of the $\mathrm{W^+W^-}$ width. The vector bosons have three polarizations and it is interesting to investigate the decays to final states with specific polarizations. The longitudinal polarizations are created by the Higgs particles and may show irregularities. For large Higgs masses the vector bosons in the final state are dominantly longitudinally polarized, which may be important for distinguishing this process from the background.

The decays into fermions or weak vector bosons proceed through tree diagrams, in contrast to the decays in groups (4) and (5), which contain loop contributions. For loop contributions one expects the branching ratios to be smaller, which is indeed the case. However, for $M_{\mathrm{H}} \leq 200\,\mathrm{GeV}$ they are still measurable because the sum over all possible particles inside the loop brings an enhancement. The gluonic decays are drowned in a huge background, in contrast to photonic decays with two stable particles, which provide a convenient way for identifying the Higgs. Both

Table 17.1. *Partial decay widths of the Higgs boson*

Higgs boson H decays into ...	Partial decay width Γ with $\lambda_i = m_i^2/m_H^2$
(1) fermions f: $H \to f\bar{f}$	$N_c \frac{G_F}{4\pi\sqrt{2}} m_f^2 m_H (1-4\lambda_f)^{3/2}$ N_c is a color factor; $N_c = \begin{cases} 1 \text{ for leptons} \\ 3 \text{ for quarks} \end{cases}$
(2) weak neutral bosons Z^0: $H \to Z^0 Z^0$	$\frac{G_F}{16\pi\sqrt{2}} m_H^3 (1-4\lambda_Z)^{1/2} (12\lambda_Z^2 - 4\lambda_Z + 1)$
(3) weak charged bosons $W^\pm$: $H \to W^+ W^-$	$\frac{G_F}{8\pi\sqrt{2}} m_H^3 (1-4\lambda_W)^{1/2} (12\lambda_W^2 - 4\lambda_W + 1)$
(4) gluons g: $H \to gg$ (via a loop containing quarks q)	$\frac{G_F}{36\pi\sqrt{2}} m_H^3 \left[\frac{\alpha_s(m_{H^0}^2)}{\pi}\right]^2 \left\vert \sum_q I_q \right\vert^2$ with $I_q = 3[2\lambda_q + \lambda_q(4\lambda_q - 1) f(\lambda_q)]$ $f(\lambda) = \begin{cases} -2\left[\sin^{-1}(1/\sqrt{4\lambda})\right]^2, & \lambda > 1/4 \\ \frac{1}{2}\left[\ln(\eta^+/\eta^-) - i\pi\right]^2, & \lambda < 1/4 \end{cases}$ $\eta^\pm = 1/2 \pm \sqrt{1/4 - \lambda}$
(5) photons γ: $H \to \gamma\gamma$	$\frac{G_F}{8\pi\sqrt{2}} m_H^3 \left(\frac{\alpha}{\pi}\right)^2 \vert I \vert^2$ with $I = \sum_q Q_q^2 I_q + \sum_l Q_l^2 I_l + I_W (+I_s)$ Q_i is the charge of particle i and $I_q = 3[2\lambda_q + \lambda_q(4\lambda_q - 1) f(\lambda_q)]$ $I_l = 2\lambda_l + \lambda_l(4\lambda_l - 1) f(\lambda_l)$ $I_W = 3\lambda_W(1 - 2\lambda_W) f(\lambda_W) - 3\lambda_W - 1/2$ $I_s = -\lambda_s[1 + 2\lambda_s f(\lambda_s)]$

photonic and gluonic modes proceed through loop diagrams; some of them are shown in Fig. 17.3 (Gunion *et al.*, 1990; Hinchlife, 1998).

The relative importance of various decays depends on the mass of the Higgs particles. We present in Table 17.2 dominant decays to various modes as a function of the Higgs mass. In the low-mass range the decay to $b\bar{b}$ pairs dominates and is identified by its decay products. The decay to γγ is less frequent ($\sim 10^{-3}$) but has a smaller background. For higher masses, decays to gauge bosons dominate.

Table 17.2. *Higgs decay modes as a function of its mass*

m_H (GeV)	Decay
90–120	$H \to b\bar{b}, \gamma\gamma$
120–140	$H \to b\bar{b}, WW^*, \tau^+\tau^-$
140–180	$H \to b\bar{b}, WW^*, ZZ^*, \tau^+\tau^-$
180–380	$H \to WW, ZZ$
>380	$H \to WW, ZZ, t\bar{t}$

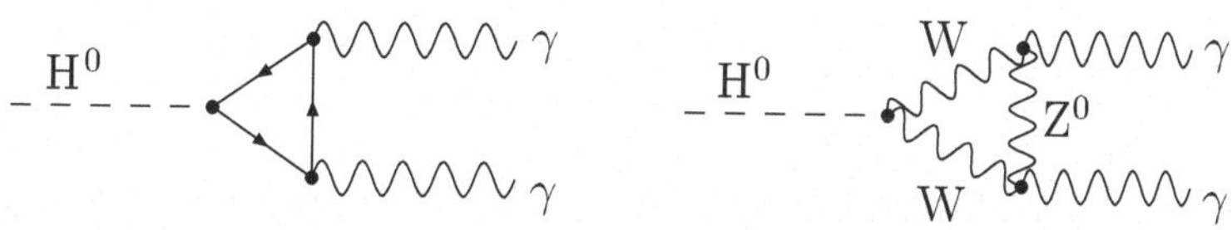

Figure 17.3. Some triangle diagrams for decay of a Higgs boson to photons.

Some decay modes may contribute even below the threshold for pair production of gauge bosons because some of them are present as virtual intermediate states. We denote them with a star as superscript; for instance W^*, which subsequently decays to $e^-\bar{\nu}$.

The calculation of decays proceeding through tree diagrams involves a matrix element and a two-body phase space, which are straightforward to calculate. We suggest some of them as exercises at the end of the chapter.

17.5 Production in electron–positron colliders

The discovery of the Higgs particle is of paramount importance. For this reason there have been large experiments designed to discover them. The early expectations of a light Higgs have been ruled out experimentally. A light Higgs appears in the decay of the known particles, like mesons, charmonium, and the gauge bosons. Intensive searches in these decays did not produce any evidence for Higgs particles. The next possibility is to produce them in electron–positron colliders through the process shown in Fig. 17.4:

$$e^+e^- \to V^* \to \underset{\hookrightarrow \text{ leptons}}{V} + H \qquad \text{with} \quad (V = \text{W}, \text{Z}) \tag{17.15}$$

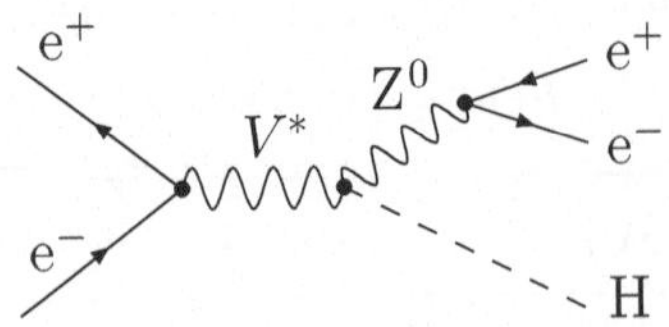

Figure 17.4. Bremsstrahlung from weak bosons.

which is known as bremsstrahlung from the weak gauge bosons. For lower energies ($\sqrt{s} \approx$ 200–500 GeV) bremsstrahlung dominates and has been searched for at the LEP collider at CERN. The cross section is maximal at center-of-mass energies close to the Z resonance, where it is given by (Gunion *et al.*, 1990)

$$\sigma(e^+e^- \to ZH) = \frac{G_F^2 M_Z^4}{96\pi}\left[1 + (1 - 4\sin^2\theta_W)^2\right]\frac{8k}{\sqrt{s}}\left[\frac{k^2 + 3M_Z^2}{\left(s - M_Z^2\right)^2 + M_Z^2\Gamma^2}\right]. \tag{17.16}$$

Here k is the center-of-mass momentum of the Z boson produced. The detection of this channel observes the outgoing Z and reconstructs the Higgs boson's invariant mass as

$$M_H^2 = s - 2\sqrt{s}E_Z + M_Z^2. \tag{17.17}$$

The searches at CERN established the bound

$$M_H > 114.4\,\text{GeV}.$$

A similar process occurs in hadron colliders, where the initial leptons are replaced by quarks to give

$$q\bar{q} \to V^* \to V + H.$$

For center-of-mass energies higher than 500 GeV, the fusion process

$$e^+e^- \to \nu\bar{\nu} + H,$$

shown in Fig. 17.5, begins to dominate. For high energies the initial partons or leptons radiate not only photons but also W and Z gauge bosons. Here one must be careful to include all possible diagrams in order to satisfy gauge invariance. The calculation follows a method very similar to that for the emission of photons, in the Weizsacker–Williams approximation, with special attention paid to the longitudinal and transverse polarization (Gunion *et al.*, 1990). In this way one derives an effective

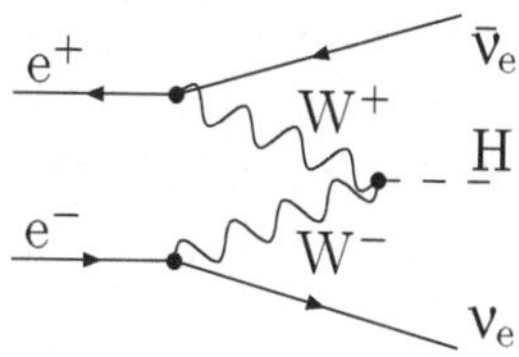

Figure 17.5. W–W fusion.

W-emission approximation, which leads to the cross section (Wilczek, 1977; Cahn and Dawson, 1984; Jones and Petkov, 1979)

$$\sigma(e^+e^- \to \nu\bar{\nu}H) \approx \frac{G_F^3 M_W^4}{4\sqrt{2}\pi^3}\left[\ln\left(\frac{s}{M_H^2}\right) - 2\right]. \tag{17.18}$$

This reaction is attractive because it has a low background. It starts smaller than the bremsstrahlung reaction in Eq. (17.15), but grows with increasing energy to become dominant at $\sqrt{s} \approx 500\,\text{GeV}$.

Finally, it is possible to create the Higgs particle in photon fusion:

$$\gamma\gamma \to H \to b\bar{b}.$$

For low-energy photons this process is very small. However, for intense high-energy laser beams it has a resonance structure and may become important. For Higgs masses larger than 150 GeV the dominant final states are two gauge bosons.

17.6 Production in hadron colliders

The high-energy electron–positron collider (LEP) completed its runs and set the bound of 114.4 GeV. The next searches will be taking place in hadron colliders. Searches are already taking place at the Tevatron and a large hadron collider (LHC) is under construction at CERN. Several of the reactions in the previous section become hadronic reactions once we replace the initial leptons by quarks or quark–antiquark pairs.

In hadronic collisions the process of gluon fusion is the most important source for Higgs-particle production. In this process the Higgs couples via a quark triangle diagram to the gluons. The point-like cross section at parton level is (Kniehl, 1994)

$$\sigma(gg \to H) = \frac{\pi^2}{8}\frac{\Gamma(H \to gg)}{M_H}\delta\left(\hat{s} - M_H^2\right). \tag{17.19}$$

For the total hadronic cross section it is very important to take higher-order gluonic corrections into account, because their contribution is positive and increases the cross section. The cross section decreases with increasing Higgs mass, although the Yukawa coupling grows with the loop-quark mass.

At the Tevatron, a $\bar{\mathrm{p}}\mathrm{p}$ collider at Fermilab, the two efficient mechanisms are gluon fusion and associated production with a W or a Z:

$$\mathrm{gg} \to \mathrm{H} \to \mathrm{b}\bar{\mathrm{b}}, \tag{17.20}$$

$$\mathrm{q}\bar{\mathrm{q}} \to \mathrm{W}^*/\mathrm{Z}^* \to \mathrm{W}/\mathrm{Z} + \mathrm{H}. \tag{17.21}$$

Although the gg process has the larger cross section, ~ 1 pb (one picobarn) at $M_{\mathrm{H}} = 115$ GeV, it is hard to detect it for $M_{\mathrm{H}} < 130$ GeV. In this mass range the dominant decay is $\mathrm{H} \to \mathrm{b}\bar{\mathrm{b}}$ (see Fig. 1 in Hinchlife (1998)), which is swamped by a multijet background. For these masses only the production of a Higgs in association with a vector boson has enough sensitivity. The WH and ZH channels give a clear signal with lepton(s), missing neutrino(s) and two b-jets. The searches continue at the Tevatron, where there are plans to reach an integrated luminosity of $\sim 8\,\mathrm{fb}^{-1}$.

The Tevatron has the potential of discovering the standard-model Higgs with masses less than ~ 170 GeV. Heavier Higgses must wait for the LHC, which will reach a center-of-mass energy of 14 TeV. It will run for an integrated luminosity of $300\,\mathrm{fb}^{-1}$ and can discover Higgs particles with masses up to $1\,\mathrm{TeV} = 10^3$ GeV. At these high energies, in addition to the gluon–gluon fusion, the quarks radiate gauge bosons. A new process of W–W fusion into Higgs, analogous to the one shown in Fig. 17.5,

$$\mathrm{W} + \mathrm{W} \to \mathrm{H} \to \gamma\gamma,\ \mathrm{ZZ}^*, \ldots \tag{17.22}$$

becomes significant. If the standard-model Higgs has a mass above twice the Z mass, the discovery will be through the channel

$$\mathrm{pp} \to \mathrm{H} + \mathrm{hadrons} \to \mathrm{ZZ} + \cdots \to (l^+l^-)(l^+l^-) + \cdots. \tag{17.23}$$

This is called the golden channel for Higgs production and decays. Both lepton pairs will have the mass of the Z boson, making possible the reduction of many backgrounds.

If the mass of the Higgs boson is much bigger than the mass of the W or Z boson, $M_{\mathrm{H}} \gg 2M_{\mathrm{W}}$, the width is

$$\Gamma(\mathrm{H} \to \mathrm{WW}) \approx \frac{3G_{\mathrm{F}}}{16\pi\sqrt{2}} M_{\mathrm{H}}^3 \approx 0.48\,\mathrm{TeV}\left(\frac{M_{\mathrm{H}}}{1\,\mathrm{TeV}}\right)^3. \tag{17.24}$$

This width grows rapidly with the mass of the Higgs particle and is very broad for large masses. The increased width of the Higgs and the reduced production rate determine the upper limit for detecting a heavy Higgs.

17.7 Other symmetry-breaking schemes

There is so much supporting evidence for a Higgs particle with a mass lower than 1 TeV that the chances of discovering it at the LHC are very high. If the Higgs is not of the simple form discussed in this chapter, there are strong arguments from unitarity and from higher-order corrections that another mechanism must be operating to reduce the fast growth of amplitudes and cross sections. Thus either the Higgs will show up as the simple scalar of the standard model, or a more complicated structure must appear.

Extensions of the standard model can have more complicated spectra of Higgs bosons. A popular extension is supersymmetry with two Higgs doublets whose neutral components have two vacuum expectation values. The physical particles are now a charged boson ($H^{\pm}$), two neutral scalar Higgses (H_1^0 and H_2^0), and one pseudoscalar (A) (Hinchlife, 1998).

Charged Higgs bosons can be pair-produced in e^+e^- and $q\bar{q}$ annihilation. The chance of detecting them depends on the energy of the collider and the branching ratios to $\nu\tau$, $c\bar{s}$, and $c\bar{b}$. Searches at the LEP did not discover them and thereby set upper limits on their masses.

In the simplest version of the supersymmetric model the mass of the lightest neutral scalar depends on the top quark's mass, the ratio of the two vacuum expectation values, and the masses of other supersymmetric particles. For $M_{\text{top}} = 175\,\text{GeV}$, there is a bound $M_{H_1^0} \lesssim 130$ GeV, provided that the ratio of the vacuum expectation values is large. This mass is within the range of the Tevatron and LHC, as we discussed earlier.

We mention a final possibility, whereby the Higgs particle is a bound state of quark–antiquark pairs. In such theories a new strong interaction must be present to bind the quark pair together. A dynamical symmetry breaking has been formulated with a top-quark condensate (Bardeen *et al.*, 1996), in analogy with the BCS theory of superconductivity. The low-energy effective theory is the standard model supplemented with relationships connecting masses of the top quark, W boson, and Higgs boson.

Another extension introduces new quarks (techniquarks) coupled to W and Z bosons and bound together by new gluons, the technigluons. There is a new coupling constant, which runs to become strong at a scale ~1 TeV. The theory is a scaled-up version of QCD with heavy fermions U and D imitating the light up and down quarks. The strong interactions at 1 TeV cause a spontaneous breaking of the $SU(2)_L \times SU(2)_R$ symmetry of the new quarks, producing heavy color scalars thc new Goldstone bosons (Fahri and Susskind, 1979). Searches for the new scalars follow the methods we described in this and the previous section.

Problems for Chapter 17

1. Show that the decay rates to transverse (±) and longitudinally (L) polarized W bosons are

$$\Gamma(\mathrm{H}^0 \to \mathrm{W}_+\mathrm{W}_+) = \Gamma(\mathrm{H}^0 \to \mathrm{W}_-\mathrm{W}_-) = \frac{g^2}{16\pi}\frac{M_\mathrm{W}^2}{M_\mathrm{H}}\left(1 - \frac{4M_\mathrm{W}^2}{M_\mathrm{H}^2}\right)^{\frac{1}{2}}$$

$$\Gamma(\mathrm{H}^0 \to \mathrm{W}_\mathrm{L}\mathrm{W}_\mathrm{L}) = \frac{g^2}{64\pi}\frac{M_\mathrm{H}^3}{M_\mathrm{W}^2}\left(1 - \frac{2M_\mathrm{W}}{M_\mathrm{H}^2}\right)^2\left(1 - \frac{4M_\mathrm{W}^2}{M_\mathrm{H}^2}\right)^{\frac{1}{2}}$$

$$\Gamma(\mathrm{H}^0 \to \mathrm{W}_\mathrm{L}\mathrm{W}_\pm) = 0.$$

Hence, for $M_\mathrm{H} \gg M_\mathrm{W}$, the W bosons from Higgs decay are dominantly longitudinally polarized.

2. Show that the amplitude for $\mathrm{W}_\mathrm{L}^+\mathrm{W}_\mathrm{L}^- \to \mathrm{W}_\mathrm{L}^+\mathrm{W}_\mathrm{L}^-$ in the limit where s, $M_\mathrm{H}^2 \gg M_\mathrm{W}^2$, M_Z^2 is

$$\mathcal{M}(\mathrm{W}_\mathrm{L}^+\mathrm{W}_\mathrm{L}^- \to \mathrm{W}_\mathrm{L}^+\mathrm{W}_\mathrm{L}^-) = -\sqrt{2}Gm_\mathrm{H}^2\left(\frac{s}{s - M_\mathrm{H}^2} + \frac{t}{t - M_\mathrm{H}^2}\right)$$

with s and t the Mandelstam variables. If $m_\mathrm{H} \to \infty$, the amplitude grows linearly in s. Calculate now the $l = 0$ partial-wave contribution

$$a_0 = \frac{1}{16\pi s}\int_{-s}^{0} \mathrm{d}t\, \mathcal{M}(\mathrm{W}_\mathrm{L}^+\mathrm{W}_\mathrm{L}^- \to \mathrm{W}_\mathrm{L}^+\mathrm{W}_\mathrm{L}^-)$$

and show that, for $s \gg M_\mathrm{H}^2$, the result for a_0 is

$$a_0 = -\frac{GM_\mathrm{H}^2}{8\pi\sqrt{2}}.$$

This result together with the unitarity condition (17.12) gives Eq. (17.13).

3. A simple estimate of the process of gluon–gluon fusion is obtained by calculating the Drell–Yan process. Consider the reaction $\mathrm{gg} \to \mathrm{H} \to \gamma\gamma$ with the mass M_H varying, i.e. the Higgs is off the mass shell. The point cross section is given in Eq. (17.19) and the decay width is listed in Table 17.1.
   (i) Identify Q^2 in Section 10.5 with M_H^2 and write the Drell–Yan cross section. For the gluon distribution function adopt the simplified form

$$f_\mathrm{g}(x) = 8\frac{1}{x}(1 - x)^7\,.$$

   Compute the production cross section in proton–proton collisions.
   (ii) Compute the cross section numerically for a Higgs of 130 GeV for pp collisions of center-of-mass energy 5–20 TeV. To simplify the calculation, you may treat $I(\lambda)$ with $\lambda = m_\mathrm{t}^2/m_\mathrm{H}^2$ as constant.

References

Bardeen, W. A., Hill, C. T., and Lindner, M. (1996), *Phys. Rev.* **D41**, 1647
Bardin, D., and Passarino, G. (1999), *The Standard Model in the Making* (Oxford, Oxford Science Publications)
Cahn, R. N., and Dawson, S. (1984), *Phys. Lett.* **B136**, 196
Erler, J., and Langacker, P. (2004), *Phys. Lett.* **B592**, 144
Fahri, E., and Susskind, L. (1979), *Phys. Rev.* **D20**, 3404
Gluck, M., Jimenez-Delgado, P., and Reya, E. (2005), *Phys. Rev. Lett.* **95**, 022002, and references therein
Gunion, J. F., Haber, H. E., Kane, G., and Dawson, S. (1990), *The Higgs Hunter's Guide* (New York, Addison-Wesley); Section 2.1 (for decays) and p. 139 (for Higgs production)
Hinchlife, I. (1998), *Eur. Phys. J.* **C3**, 1
Kniehl, B. (1994), *Phys. Rep.* **240**, 211
Lee, B. W., Quigg, C., and Thacker, G. B. (1977a), *Phys. Rev. Lett.* **38**, 833
(1977b), *Phys. Rev.* **D16**, 1519
Linde, A. D. (1976), *JETP Lett.* **23**, 64
Jones, D. R. T., and Petkov, S. T. (1979), *Phys. Lett.* **B84**, 440
Wilczek, F. (1977), *Phys. Rev. Lett.* **39**, 1304

Select bibliography

Roth, S. (2007), *Precision Electroweak Physics at Electron–Positron Colliders* (Berlin, Springer-Verlag)

Epilogue

In this book, the reader has had the opportunity to follow the development of the electroweak theory and the discovery of several new phenomena predicted by the theory. We have analyzed several of them in various chapters. The richness of the field and its high level of accuracy have been achieved with the help of several very large, very accurate, and refined experiments.

The potential for discovery has not yet been exhausted because the theory must be completed with the discovery of the Higgs particle(s) or some other symmetry-breaking scheme. Experiments at the LHC will either discover the Higgs boson or find new interactions indicating another mechanism for the breaking of symmetry. There is another observation in addition to symmetry-breaking that demands extension of the theory: the mixing and mass differences of neutrinos. Most of our colleagues speculate about a larger theory. Grand unified theories will unify the electroweak with the strong interactions, at some high energy bringing the three coupling constants together. The main issue here is to find predictions unique to the grand unified theory that will be verified by experiment. The symmetry of the new theory and the particle classification within the group remain open issues.

An alternative theory is supersymmetry, with many more particles. Supersymmetry is a symmetry relating fermions to bosons and must be broken. There are many different symmetry-breaking schemes. Predictions of the minimal supersymmetric theory will be tested at the LHC.

A parallel development has been the association of weak and electromagnetic interactions with astronomical and cosmological phenomena. For instance, particle interactions determine the spectrum of the cosmic background radiation. Furthermore, with small extensions, the present theory explains the generation of baryons. One attractive scenario considers the generation of a lepton asymmetry converted into a baryon asymmetry at the energy scale of the electroweak phase transition ($v = 246$ GeV), generating an excess of matter. The interplay between gravity and particle interactions is responsible for the formation of large structures. Such developments have brought about a closer cooperation between particle physicists and cosmologists.

I was tempted to include possible new developments that would lead to diverse predictions. This is a difficult task, where past experience indicates that Nature frequently selects solutions beyond our imagination. I resisted this temptation, and decided to stop here with an open-ended conclusion that invites future thought.

Appendix A

Conventions, spinors, and currents

A.1 Conventions

The space-time coordinates $(t, x, y, z) = (t, \vec{x})$ are denoted by a contravariant four-vector (c and $\hbar$ are set equal to 1):

$$x^\mu = (x^0, x^1, x^2, x^3) = (t, x, y, z). \tag{A.1}$$

The metric tensor is

$$g_{\mu\nu} = g^{\mu\nu} = \begin{pmatrix} 1 & 0 & 0 & 0 \\ 0 & -1 & 0 & 0 \\ 0 & 0 & -1 & 0 \\ 0 & 0 & 0 & -1 \end{pmatrix}, \tag{A.2}$$

$$p^\mu = (p_0, \vec{p}), \quad p_\mu = g_{\mu\nu} p^\nu = (p_0, -\vec{p}\,). \tag{A.3}$$

Momentum four-vectors are similarly defined,

$$p^\mu = (E, p_x, p_y, p_z), \tag{A.4}$$

and the inner product

$$p_1 \cdot p_2 = p_{1\mu} p_2^\mu = (E_1 E_2 - \vec{p}_1 \vec{p}_2). \tag{A.5}$$

We frequently meet products of the totally antisymmetric tensor $\varepsilon_{\alpha\beta\gamma\mu}$ (note $g_\mu^\nu = \delta_\mu^\nu$)

$$\varepsilon_{\alpha\beta\gamma\mu}\varepsilon^{\alpha\beta\gamma\nu} = -6\delta_\mu{}^\nu, \tag{A.6}$$

$$\varepsilon_{\alpha\beta\mu\nu}\varepsilon^{\alpha\beta\rho\sigma} = -2\begin{vmatrix} \delta_\mu^\rho & \delta_\nu^\rho \\ \delta_\mu^\sigma & \delta_\nu^\sigma \end{vmatrix}, \tag{A.7}$$

$$\varepsilon_{\alpha\mu\nu\sigma}\varepsilon^{\alpha\lambda\rho\tau} = \begin{vmatrix} \delta_\mu^\lambda & \delta_\nu^\lambda & \delta_\sigma^\lambda \\ \delta_\mu^\rho & \delta_\nu^\rho & \delta_\sigma^\rho \\ \delta_\mu^\tau & \delta_\nu^\tau & \delta_\sigma^\tau \end{vmatrix}. \tag{A.8}$$

A.2 Dirac matrices and spinors

Anticommutation of γ-matrices:

$$\{\gamma^\mu, \gamma^\nu\} = \gamma^\mu\gamma^\nu + \gamma^\nu\gamma^\mu = 2g^{\mu\nu}, \tag{A.9}$$

$$\gamma^5 \equiv \mathrm{i}\gamma^0\gamma^1\gamma^2\gamma^3, \quad \{\gamma^\mu, \gamma^5\} = 0. \tag{A.10}$$

The σ-matrix:

$$\sigma^{\mu\nu} = \frac{\mathrm{i}}{2}[\gamma^\mu, \gamma^\nu]. \tag{A.11}$$

Reduction of the product of three γ-matrices:

$$\gamma^\mu\gamma^\rho\gamma^\nu = S^{\mu\rho\nu} + \mathrm{i}\epsilon_\lambda^{\mu\nu\rho}\gamma^\lambda\gamma_5, \tag{A.12}$$

with

$$S^{\mu\rho\nu} = g^{\mu\rho}\gamma^\nu + g^{\rho\nu}\gamma^\mu - g^{\mu\nu}\gamma^\rho. \tag{A.13}$$

A familiar representation of γ-matrices is

$$\gamma^0 = \begin{bmatrix} \mathbf{1} & 0 \\ 0 & -\mathbf{1} \end{bmatrix}, \tag{A.14}$$

$$\{\gamma^i\} = \boldsymbol{\gamma} = \begin{bmatrix} 0 & \boldsymbol{\sigma} \\ -\boldsymbol{\sigma} & 0 \end{bmatrix}, \quad \gamma_5 = \gamma^5 = \begin{bmatrix} 0 & \mathbf{1} \\ \mathbf{1} & 0 \end{bmatrix}, \tag{A.15}$$

where

$$\boldsymbol{\sigma}^1 = \begin{bmatrix} 0 & 1 \\ 1 & 0 \end{bmatrix}, \quad \boldsymbol{\sigma}^2 = \begin{bmatrix} 0 & -\mathrm{i} \\ \mathrm{i} & 0 \end{bmatrix}, \quad \boldsymbol{\sigma}^3 = \begin{bmatrix} 1 & 0 \\ 0 & -1 \end{bmatrix} \tag{A.16}$$

are the familiar Pauli matrices and

$$\mathbf{1} = \begin{bmatrix} 1 & 0 \\ 0 & 1 \end{bmatrix}$$

is the 2×2 unit matrix.

The spinors u and v satisfy the Dirac equation,

$$(\not{p} - m)u(p, s) = 0, \tag{A.17}$$

$$(\not{p} + m)v(p, s) = 0. \tag{A.18}$$

The normalization of spinors is

$$\bar{u}(p, s)u(p, s) = 2m, \tag{A.19}$$

$$\bar{v}(p, s)v(p, s) = -2m, \tag{A.20}$$

and the completeness relation is

$$\sum_s u(p, s)\bar{u}(p, s) = \not{p} + m, \tag{A.21}$$

$$\sum_s v(p, s)\bar{v}(p, s) = \not{p} - m. \tag{A.22}$$

A.3 Currents

Vector:

$$J_\mu(x) = \bar{\Psi}(x)\gamma_\mu\Psi(x) = \Psi(x)^+\gamma_0\gamma_\mu\Psi(x). \tag{A.23}$$

Axial:

$$J_{\mu 5}(x) = \bar{\Psi}(x)\gamma_\mu\gamma_5\Psi(x). \tag{A.24}$$

Decompositions of the currents or products of them are very useful. Let $\ell_\mu = p_\mu + p'_\mu$ and $q_\mu = p'_\mu - p_\mu$, then

$$\bar{u}(p')\gamma^\mu u(p) = \frac{1}{2m}\bar{u}(p')(\ell^\mu + \mathrm{i}\sigma^{\mu\nu}q_\nu)u(p), \tag{A.25}$$

$$\bar{u}(p')\gamma^\mu\gamma_5 u(p) = \frac{1}{2m}\bar{u}(p')(\gamma_5 q^\mu + \mathrm{i}\gamma_5\sigma^{\mu\nu}\ell_\nu)u(p), \tag{A.26}$$

$$\bar{u}(p')\mathrm{i}\sigma^{\mu\nu}\ell_\nu u(p) = -\bar{u}(p')q^\mu u(p), \tag{A.27}$$

$$\bar{u}(p')\mathrm{i}\sigma^{\mu\nu}q_\nu u(p) = \bar{u}(p')(2m\gamma^\mu - \ell^\mu)u(p). \tag{A.28}$$

Additional identities can be found in Appendix A of the article by M. Nowakowski, E. Paschos and J. M. Rodriguez (*Eur. J. Phys.* **26**, 545–560, 2005) and in Appendix C of this book.

Appendix B

Cross sections and traces

B.1 Cross sections

The matrix element $\mathcal{M}$ appearing in cross sections and decay rates is Lorentz-invariant and dimensionless. The expression for the cross section is

$$d\sigma = \frac{1}{|\vec{v}_1 - \vec{v}_2|} \frac{1}{2E_{p_1}} \frac{1}{2E_{p_2}} |\mathcal{M}|^2 (2\pi)^4 \delta^4 \left(p_1 + p_2 - \sum_{i=1}^{n} k_i \right) \frac{d^3 k_1}{2E_1 (2\pi)^3} \cdots \frac{d^3 k_n}{2E_n (2\pi)^3} s. \tag{B.1}$$

The flux factor is frequently computed in the laboratory frame or the center-of-mass frame. In general,

$$\frac{1}{|\vec{v}_1 - \vec{v}_2|} = \frac{mM}{\left[(p_1 \cdot p_2)^2 - m^2 M^2\right]^{1/2}}. \tag{B.2}$$

The factor s is

$$s = \prod_i \frac{1}{k_i!} \tag{B.3}$$

if there are k_i identical particles of species i in the final state.

The decay width for a particle moving with energy E is

$$d\Gamma = \frac{1}{2E} |\mathcal{M}|^2 (2\pi)^4 \delta^4 \left(p - \sum_{i=1}^{n} k_i \right) \frac{d^3 k_1}{2E_1 (2\pi)^3} \cdots \frac{d^3 k_n}{2E_n (2\pi)^3} s. \tag{B.4}$$

B.2 Contraction identities and traces

$$\not{a}\not{b} = 2a \cdot b - \not{b}\not{a}, \tag{B.5}$$

$$\gamma^\lambda \gamma_\lambda = 4, \tag{B.6}$$

$$\gamma^\lambda \gamma^\mu \gamma_\lambda = -2\gamma^\mu, \tag{B.7}$$

$$\gamma^\lambda \gamma^\mu \gamma^\nu \gamma_\lambda = 4g^{\mu\nu}, \tag{B.8}$$

$$\gamma^\lambda \gamma^\mu \gamma^\nu \gamma^\rho \gamma_\lambda = -2\gamma^\rho \gamma^\nu \gamma^\mu, \tag{B.9}$$

$$\gamma^\lambda \gamma^\mu \gamma^\nu \gamma^\rho \gamma^\sigma \gamma_\lambda = 2(\gamma^\sigma \gamma^\mu \gamma^\nu \gamma^\rho + \gamma^\rho \gamma^\nu \gamma^\mu \gamma^\sigma), \tag{B.10}$$

$$\gamma^\lambda \sigma^{\mu\nu} \gamma_\lambda = 0, \tag{B.11}$$

$$\gamma^\lambda \sigma^{\mu\nu} \gamma^\rho \gamma_\lambda = 2\gamma^\rho \sigma^{\mu\nu}. \tag{B.12}$$

The trace of an odd product of γ^μ-matrices vanishes:

$$\mathrm{Tr}(\gamma^5) = 0 \tag{B.13}$$

$$\mathrm{Tr}(\gamma^\mu\gamma^\nu) = 4g^{\mu\nu} \tag{B.14}$$

$$\mathrm{Tr}(\sigma^{\mu\nu}) = 0 \tag{B.15}$$

$$\mathrm{Tr}(\gamma^\mu\gamma^\nu\gamma^5) = 0 \tag{B.16}$$

$$\mathrm{Tr}(\gamma^\mu\gamma^\nu\gamma^\rho\gamma^\sigma) = 4(g^{\mu\nu}g^{\rho\sigma} - g^{\mu\rho}g^{\nu\sigma} + g^{\mu\sigma}g^{\nu\rho}) \tag{B.17}$$

$$\mathrm{Tr}(\gamma^5\gamma^\mu\gamma^\nu\gamma^\rho\gamma^\sigma) = -4\mathrm{i}\varepsilon^{\mu\nu\rho\sigma} = 4\mathrm{i}\varepsilon_{\mu\nu\rho\sigma} \tag{B.18}$$

$$\mathrm{Tr}(\not{a}_1\not{a}_2\ldots\not{a}_{2n}) = \mathrm{Tr}(\not{a}_{2n}\ldots\not{a}_2\not{a}_1) \tag{B.19}$$

$$\begin{aligned}\mathrm{Tr}(\not{a}_1\not{a}_2\ldots\not{a}_{2n}) &= a_1\cdot a_2\,\mathrm{Tr}(\not{a}_3\ldots\not{a}_{2n}) - a_1\cdot a_3\,\mathrm{Tr}(\not{a}_2\not{a}_4\ldots\not{a}_{2n}) + \cdots \\ &\quad + a_1\cdot a_{2n}\,\mathrm{Tr}(\not{a}_1\ldots\not{a}_{2n-1}) \\ &= 4\sum \varepsilon(a_{i_1}\cdot a_{j_1})\ldots(a_{i_n}\cdot a_{j_n}).\end{aligned} \tag{B.20}$$

ε is the signature of the permuatation $i_1 j_1 \ldots i_n j_n$ and the sum runs over the $(2n)!/(2^n n!)$ different pairings satisfying $1 = i_1 < i_2 < \cdots i_n$, $i_k < j_k$.

B.3 Some Feynman rules

In the text we gave Feynman rules for several vertices. We present here additional rules for vertices of gauge bosons:

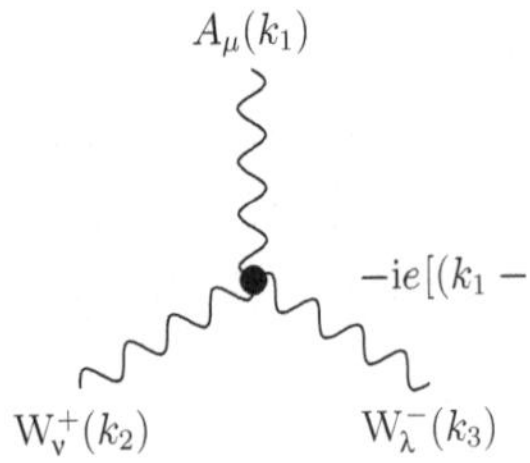

$$-\mathrm{i}e[(k_1-k_2)_\lambda g_{\mu\nu} + (k_2-k_3)_\mu g_{\nu\lambda} + (k_3-k_1)_\nu g_{\lambda\mu}]$$

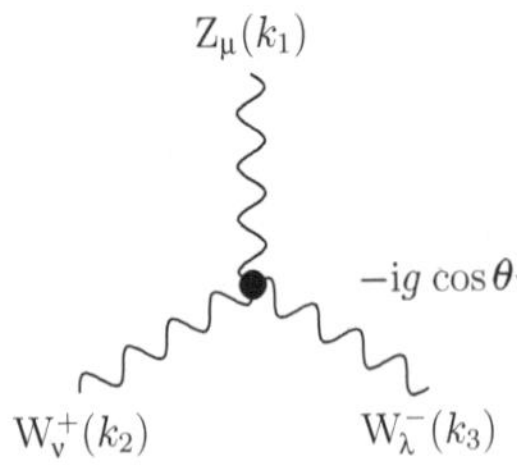

$$-\mathrm{i}g\cos\theta_W[(k_1-k_2)_\lambda g_{\mu\nu} + (k_2-k_3)_\mu g_{\nu\lambda} + (k_3-k_1)_\nu g_{\lambda\mu}]$$

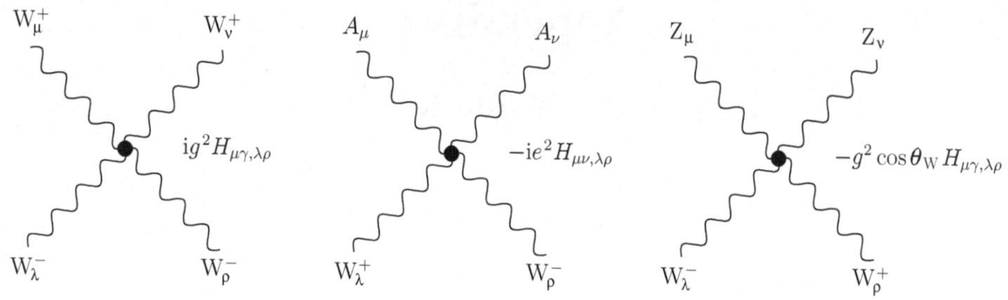

with $H_{\mu\nu,\lambda\rho} = 2g_{\mu\nu}g_{\lambda\rho} - g_{\mu\lambda}g_{\nu\rho} - g_{\mu\rho}g_{\nu\lambda}$. In graphs all momenta are taken to be entering *into* the vertices.

Appendix C

Identities for quark bilinears

There are four identities for the product of two quark bilinears:

$$\big[\bar{s}\gamma^{\mu}\gamma^{\rho}\gamma^{\nu}\gamma_{\pm}d\big]\big[\bar{q}_i\gamma_{\mu}\gamma^{\tau}\gamma_{\nu}\gamma_{\pm}q_j\big] = 4g^{\rho\tau}\big[\bar{s}\gamma^{\lambda}\gamma_{\pm}d\big]\big[\bar{q}_i\gamma_{\lambda}\gamma_{\pm}q_j\big], \tag{C.1}$$

$$\big[\bar{s}\gamma^{\mu}\gamma^{\rho}\gamma^{\nu}\gamma_{\pm}d\big]\big[\bar{q}_i\gamma_{\mu}\gamma^{\tau}\gamma_{\nu}\gamma_{\mp}q_j\big] = 4\big[\bar{s}\gamma^{\tau}\gamma_{\pm}d\big]\big[\bar{q}_i\gamma^{\rho}\gamma_{\mp}q_j\big], \tag{C.2}$$

$$\big[\bar{s}\gamma^{\nu}\gamma^{\rho}\gamma^{\mu}\gamma_{\pm}d\big]\big[\bar{q}_i\gamma_{\mu}\gamma^{\tau}\gamma_{\nu}\gamma_{\pm}q_j\big] = 4\big[\bar{s}\gamma^{\tau}\gamma_{\pm}d\big]\big[\bar{q}_i\gamma^{\rho}\gamma_{\pm}q_j\big], \tag{C.3}$$

$$\big[\bar{s}\gamma^{\nu}\gamma^{\rho}\gamma^{\mu}\gamma_{\pm}d\big]\big[\bar{q}_i\gamma_{\mu}\gamma^{\tau}\gamma_{\nu}\gamma_{\mp}q_j\big] = 4g^{\rho\tau}\big[\bar{s}\gamma^{\lambda}\gamma_{\pm}d\big]\big[\bar{q}_i\gamma_{\lambda}\gamma_{\mp}q_j\big]. \tag{C.4}$$

The derivation follows from the identity

$$\gamma^{\mu}\gamma^{\lambda}\gamma^{\nu} = g^{\mu\lambda}\gamma^{\nu} + g^{\lambda\nu}\gamma^{\mu} - g^{\mu\nu}\gamma^{\lambda} + \mathrm{i}\varepsilon^{\mu\lambda\nu\rho}\gamma_{\rho}\gamma_5 \tag{C.5}$$

and can be found in A. Sirlin, *Nucl. Phys.* B **192**, 93 (1981).

Index

9 789360 846626

Printed at Repro India Ltd.